'Looking Back

Stories of Real American Pioneers

by

Bert Entwistle

Tim —

Many Thanks for everything + for such a great publication!!

Bert

blackmulepress.com

Books by Bert Entwistle

The Drift

Uranium Drive-In

The Black Rose Banker

The Taylor Legacy

New Mexico

Murder in the Dell

Leftover Soldiers

'Lookin Back'

Author's Note:

This book is a collection of 50 magazine columns I originally wrote and published in **Working Ranch Magazine,** under the title of **'*Looking Back.*'** All the charactors are real people that helped make our country what it is today. These stories of their explorations, adventures, businesses, and lives have been gleaned from their first-person writings, other period accounts, and hundreds of hours of research. The romance of the books, movies, and television shows I grew up with is a long way from the reality of the times. The men and women in these stories were good guys and bad guys and sometimes they were both, but the impact they had on this country and our lives is enormous.

As the country began to move west, people who were Americans by birth and a flood of immigrants from every corner of the world, began the great migration to the wild lands of America. The diversity of this movement was incredible. Men, women and children of every ethnicity, color, religion, and age pushed west looking for a new life. By the end of the Civil War they came by covered wagon and horseback, and even more walked the trails. The gates to the future had been thrown open and opportunity was calling. These stories are about their successes and failures as well as what it was that drove them and the legacy they left us.

Bert Entwistle, Colorado Springs, Colorado, 2019

All rights reserved. No portion of this book may be reproduced without written permission from the publisher except by a reviewer who may quote brief passages in a review.

Published and printed in the United States of America.

© 2019 Black Mule Press - Colorado Springs, Colorado

First Edition: July, 2019

ISBN: 978-0-9896761-9-9

Library of Congress Control Number: 2019905446

PRICE: $16.00

Website: blackmulepress.com

Front cover art: 'Crystal Mill' © Bert Entwistle

Back cover art: 'Heading Down' © Bert Entwistle

Author photo: © Jeremy Entwistle

blackmulepress.com

About Working Ranch Magazine

Years ago, in Amarillo, Texas, I met a young Irishman named Tim O'Byrne at a rodeo where I was working as the event photographer. He enthusiastically showed me a copy of his new project called *Working Ranch Magazine*. We'd never heard of each other before, but after a few minutes of conversation, I asked him if he could use another writer or photographer. Being an Englishman myself, it seemed like we were destined to meet. After a few minutes, we shook hands and I have been writing feature stories, columns, and providing photographs ever since. All these years later, *Working Ranch Magazine* is now the premier ranching magazine in the business today and I'm proud to have been with it since the beginning.

Many thanks to my good friend Tim O'Byrne, publisher of that great magazine for the chance to be part of something so cool!

Table of Contents:

Manifest Destiny - 1

The Log of a Cowboy - 5

A Life Well Lived - 9

Boone's Road - 13

Splendid Behavior - 17

Burnin' Hair - 21

American Bred - 25

"8 Cows and One Stud Bull" - 29

Canyons, Cows and Six-Guns - 33

Oklahoma City's First Citizen - 37

Confederate Cow Cavalry - 41

Feed Lots & Cowboy Brogue - 45

Waiting for a Chinook - 49

Long Island Aged Cattle - 53

Ghosts of the Wild West - 57

The Fabulous 101 - 61

Ghosts of the Seeds-Kee-Dee - 65

The 200 Hundred Year Range War - 69

Hooves, Horns and Grass – Part-1 - 73

Hooves, Horns and Grass – Part-2 - 77

The Wild Owyhee - 81

Kittitas County - 85

The Last Squeal - 89

Highway of Hooves - 93

Mesa Verde's Cattle - 97

One Mans Legacy - 101

Painted History - 105

The Queens Calf - 109

The Genuine Real McCoy - 113

Boss of the Plains - 117

Theodore's Cattle - 121

Emporer of the Badlands - 125

The Lincoln County Cattle Queen -129

The Wild Corner - 133

The Queen City Throws a Party - 137

U.S. Forest Service, 110 Years - 141

The Texas Cattle Queen - 145

The Willamette Cattle Company - 149

The Wiregrass Herd - 153

From Butchers to Barons - 157

Uncle Dick & Charlie - 161

Kansas City Here They Come - 165

A Town Named Wibaux - 169

N.K. Boswell - 173

Nat and Deadwood - 177

Charles Littlepage Ballard - 181

"We Pointed Them North" - 185

The Fever – 189

Utah's Last Cattle King -193

The Unassigned Land - 197

Manifest Destiny

"Coming Across The Plain" C.M. Russell, 1903

In the July-August 1845 issue of the United States Magazine and Democratic Review, the editor, John O'Sullivan, published an essay he titled Annexation. It was written as a call to the U.S. Congress to annex Texas into the United States. They had voted earlier to take them in, but Texas had yet to agree. The move was a contentious issue and the

Senate was concerned about the possibilities of a war with Mexico and the expansion of another slave state. In his essay he said,

*"It is now time for the opposition to the annexation of Texas to cease. It is our '**Manifest Destiny**' to overspread the continent allotted by providence (God) for the free development of our yearly multiplying millions."*

The second time he used the phrase in his writing, it appeared in the New York Morning News on December 27, 1845, in addressing a boundary dispute in the Oregon country against Great Britain:

*". . . and that claim is by the right of our **Manifest Destiny** to overspread and to possess the whole of the continent which providence has given us."*

Although the sentiment was not new, the phrase took hold in the American mind and was picked up as the rallying call for those wanting their own piece of the West. The phrase was never mentioned in law, but it somehow gave Americans of the day the justification and the extra shot of courage needed to push on. Many Americans of the period believed that they were special, even divinely ordained, and it was not only their right, but their mission to move west. In many cases, they wanted to remake the new lands into a copy of the agrarian life they knew in the East. President James K. Polk adopted it as his philosophy, and his became one of the most expansionist administrations in history.

The idea of manifest destiny was not universally embraced by everyone, including prominent Americans like Abraham Lincoln and Ulysses Grant. The issue proved to be a bitter political and social divide. One side saw the issue as their natural right to conquest based on perceived racial, cultural, and religious superiority. The critics saw expansionism as an excuse to start wars, expand slavery, remove the Indians, and justify an unprecedented land grab.

On December 29, 1845, the annexation of Texas as the twenty-eighth state was completed. Since Mexico still considered it part of their territory in 1846, America justified going to war with Mexico over possession of Texas. After two years of war, the Americans forced Mexico to sign the Treaty of Hidalgo. Mexico had to give up its claim to its properties in New Mexico, California, and Texas for fifteen million dollars and the United States forgave more than three million dollars in Mexican debt. The Rio Grande became the permanent border with Texas.

By 1846, the number of immigrants on the Oregon Trail began to increase at a dramatic rate. From the popular starting point in Independence, Missouri, the 2,200-mile trail was flooded by thousands of travelers looking for their place in the new Oregon country. Among these early Americans finding their way west were the Mormons, arriving for the first time in the Salt Lake Valley on July 24, 1847. As soon as the first immigrants arrived, they began in earnest to settle the soon to be Utah Territory.

In 1848 gold was discovered in California and again in Colorado in 1858 adding even more people to the frontier. The rush of gold seekers to the Pikes Peak gold fields broke the Fort Laramie Treaty of 1851 and the government chose to ignore the problem.

The old Santa Fe Trail, also starting in Missouri and in use since 1821, now found itself a major highway for those looking for a new life in Southern California. Tens of thousands of immigrants took the nine-hundred-mile trip to the frontier town of Santa Fe. Once resupplied, they continued their journey west.

The idea of manifest destiny was by now so firmly entrenched in the American psyche that the flood of humanity heading west was unstoppable. In the rush to populate what many considered empty land, the predictions of the anti-expansion groups began to come true. The so-called empty land contained tens of thousands of Native Americans from dozens of different nations. The removal of the Indians from the Great Plains and the mountains had begun.

On May 20, 1862, President Lincoln signed the Homestead Act giving people a chance to own one-hundred and sixty acres if they lived on it for five years and proved it was a working homestead. The drive to fulfill our manifest destiny was being severely tested. Between the time the Homestead Act was passed, and the late 1870s, the battles between the government and the Indians continued with hundreds of lives lost on both sides.

As the push West continued, the immigrants began to realize that if the prairies held flourishing herds of buffalo, they would also hold large herds of cattle. This was the start of a new industry and settlers began to stake out ground for new farms and ranches in the lands previously promised to the Indians by treaty.

The expansion into the forbidden lands played right into the hands of the government. They wanted a way to get use of the land and remove the Indian population for good. They endorsed the slaughter of the buffalo as a way to open the prairie to the cattle and farming business. They also knew that without the buffalo, the plains nations would lose their main source of food. President Grant was pressured to kill the buffalo as a way to eliminate the *Indian problem* as the newspapers of the day called it.

By the late 1880s, most of the buffalo had been killed and the Indians were being relocated to government run reservations. The concept of manifest destiny was in full swing and the Indian land had been opened for ranching and farming. Even as the West was becoming tamer and more civilized, many weren't satisfied with what they had. In 1889, the Johnson County, Wyoming, range wars erupted between ranchers that ran their stock on the open range and the farmers and settlers who wanted their smaller properties fenced. Both were looking for their own version of manifest destiny.

In 1889 the government opened 2,000,000 acres of land known as the Oklahoma/Indian Territory to development. In one day, 50,000 settlers raced west for their piece of the pie. The West was now open — manifest destiny had done its job.

The Log of a Cowboy

"The Herd Quitter" C.M. Russell, 1897

In 1903 Andy Adams published a book called, *The Log of a Cowboy*. Although sold as a fiction novel, it was really Adam's first-person account of his years as a working cowboy. The book tells of one of his trail drives, a five-month trip driving a 3,000 head mixed herd from Brownsville, Texas, to the Blackfoot Indian agency in Northwest Montana. The outfit left Brownsville on April 1, 1882 and arrived at the Indian agency on September 8th, a journey of more than 2,100 miles and more than five months.

Born in Indiana in 1859 to a pioneer family, his father, a Confederate soldier, moved to San Antonio about 1880 looking for a fresh start among his old comrades. Always fascinated by horses and the cattle business, the fifteen-year-old Adams left home to make a life as a cowboy.

"My worst trouble was getting away from home on the morning of starting," he recounts in the book. *"Mother and my sisters, of course, shed a few tears; but my father, stern and unbending in his manner, gave me his benediction in these words: "Thomas Moore, you're the third son to leave our roof, but your father's blessing goes with you. I left my own home beyond the sea before I was your age." And as they all stood at the gate, I climbed into my saddle and rode away, with a lump in my throat which left me speechless to reply."*

By 1882, he was working as a hired hand helping to put together a herd from many different sources, including Texas and Mexican cattle. Before heading north, they planted the circle-dot brand on their left hip.

For more than ten years, Adams worked as a cowboy and helped to move the cattle from Texas to the end of the next trail. In the early '90s he decided to try his hand at business and gold mining in places like Nevada and Cripple Creek, Colorado. Both ventures failed. By 1894, he had settled in Colorado Springs, with no gold and very little else to show for his forty-three years of life.

Adams was always frustrated with what he read in the books and magazines of the period, often complaining to anyone who would listen about how inaccurate the accounts were. It was easy to see the authors had never been up close to horses and cattle. After some encouragement from friends, he began to sell articles to various publications in Colorado and eventually to the east coast magazines.

In time, he realized that the public had an appetite for western stories and began his first and most successful book, *The Log of a Cowboy*, published in 1903. The book was really a way for Adams to respond to the bad western writing of the time, and cemented his career as a novelist. Today, a hundred and sixteen years later, it is still in print and considered by most western historians to be the finest account of the late nineteenth century working cowboy ever put in words.

The book puts the reader in the saddle with the cowboy as he eats the dust of 3,000 cattle seven days a week for more than five months. Adams learned the craft as he went from the old timers as they trailed the herd north.

"On April 1, 1882, our Circle Dot herd started on its long tramp to the Blackfoot Agency in Montana. With six men on each side, and the herd strung out for three quarters of a mile, it could only be compared to some mythical serpent or Chinese dragon, as it moved forward on its sinuous, snail like course."

At the start of the drive, his boss gave everyone the advice that all cowboys learn fast on the trail:

"Boys, the secret of trailing cattle is never to let your herd know that they are under restraint. Let everything that is done be done voluntarily by the cattle. From the moment you let them off the bed in the morning until they are bedded at night, never let a cow take a step, except in the direction of its destination."

Adams wrote about choosing his remuda of six to eight horses and the saddles and tack of the period in great detail. The writing often contradicted what was being printed at the time, but his readers came to realize his articles and books were a result of his real life experiences and a sharp eye for detail. Stories of swimming the herd across rivers and dry drives and lightening induced runs (stampedes) filled the pages with the harsh reality of the trail. Encounters with bears, wild cattle, and lost cows from other herds were common.

The importance of camaraderie and humor between the cowboys was also a large part of the story. He wrote of one of the hands named Ash Borrowstone getting disturbed at night by a couple of coyotes:

"There was no more danger of attack from these cowards than from field mice, but their presence annoyed Ash, and as he dared not shoot, he threw his boots at the varmints. Imagine to his chagrin the next morning to find that one boot had landed among the banked embers of the camp-fire and was burned to a crisp."

On final delivery to the Blackfoot Agency, he encountered several local Indians:

"The next morning, before we reached the agency, a number of gaudily bedecked bucks and squaws rode out to meet us. Physically, they were fine specimens of the aborigines."

The stories of the Indians were like icing on the cake for fans of Adams' work. From his Colorado Springs home, he produced another six books after *Log of a Cowboy* and many more articles. He wrote for a wide variety of audiences including young readers and those who love short stories. As he got older, he was known for encouraging and sponsoring young western writers in authentic western fiction. Adams ran unsuccessfully three times for El Paso County (Colorado) sheriff and lived quietly as a bachelor until his death on September 26, 1935.

For lovers of Western history, a night with Andy Adams on the long trail drive north will make you feel like you're in the saddle riding night herd under a starlit sky to the music of the coyotes.

A Life Well Lived

Edward Beale's Camel Corps. 1858 - 1859

How much adventure can one man pack into seventy-one years? If you were Edward Beale, the answer is, enough for a bunch of people — a really big bunch. Ned, as his friends called him, was born into an adventurous life on February 4, 1822. His father, George, won the Congressional Medal for Valor for his service in the War of 1812. A smart and ambitious young man, he was appointed to the United States Naval School in

Philadelphia by President Andrew Jackson, graduating in 1842. During those years he sailed to Brazil, Russia, the West Indies, and then to South America and Europe.

In 1845 he sailed under Robert F. Stockton on the *Congress* to Texas, Oregon, and California. Coming upon a Danish ship, the Americans captured it and Beale was instructed to sail it to England, disguise his identity, and act as a spy to learn what England was thinking on the Oregon boundary dispute. Returning to Washington, he reported to President Polk that the British had been making ready for war.

For his extraordinary service, Beale was promoted, and given a packet of sensitive government documents to deliver to Navy Secretary Bancroft in Peru. Landing on the Gulf side of Panama, he crossed the isthmus on mules and small boats. From there he sailed to Callo, Peru and delivered his valuable package. From Peru, they sailed the *Congress* to Honolulu then on to Monterey, California.

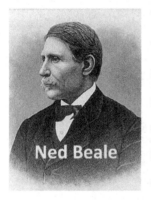

The Mexican American war was well under way when they reached California, and Stockton gave Beale a new assignment; serve with the land forces already engaged with the Mexican Army. On December 6, 1846, while fighting for General Stephen Kearny, they became surrounded by Mexican troops in the famous *Battle of San Pasqual.* Beale, his Delaware Indian servant, and Kit Carson crawled through enemy lines in the dark of night and made their way to San Diego. They gathered reinforcements and returned to rout the Mexicans and save the detachment.

In February, Stockton sent him to Washington DC, with dispatches for the president. While in Washington he appeared as a witness in the infamous trial of 'Pathfinder' John C. Fremont. Beale made at least six more cross country trips for Stockton and Washington. On his second trip he covertly crossed through Mexico in disguise to deliver proof of the California gold strike to Washington. While in Washington on his fourth trip, he married Mary Edwards, daughter of a prominent Pennsylvania Representative.

Resigning from the Navy, he returned to California as a manager for Stockton's growing land holdings. By 1853, President Fillmore needed a Superintendent of Indian Affairs for California and Nevada and tapped Beale for the job.

Leaving Washington with a party of 13 men, he was tasked with finding a possible route for a transcontinental railroad. Travelling through southern Colorado and Utah, he surveyed and documented possible routes, reaching Los Angeles in August. California Governor, John Bigler, then appointed him brigadier general in the state militia.

By the late 1850s, Americans and new immigrants were pushing west, and the country between the east coast and California contained very few passable roads. Because of his past experience, President Buchanan appointed him to survey a wagon road to connect Fort Defiance, on the New Mexico/Arizona border, to the Colorado River on the California/Arizona border. The road crossed the river near present-day Needles.

This was a wild stretch of desert, and Beale agreed to an experiment suggested by Secretary of War Jefferson Davis several years earlier — camels. Beal's Camel Corps. used animals imported from Tunisia as pack animals in this expedition and again in 1858 and 1859 on a survey for a section of road from Fort Smith, Arkansas, to the Colorado River. The camels proved to be perfect for the desert work, carrying heavier loads and needing less water and feed. However, the Army was forced to drop the Camel Corps. when they realized the horses and mules were frightened whenever they came near.

The route of the Beale Wagon Road was immediately popular with travelers and immigrants, and was eventually used by the famous Route 66, the Santa Fe Railway and interstate 40. Beale later wrote:

". . . It is the shortest route from our western frontier by 300 miles, being nearly directly west. It is the most level, our wagons only double teaming only once in the entire distance,

and that a short hill, and over a surface heretofore unbroken by wheels or trail of any kind."

In 1861, President Lincoln appointed Beale Surveyor General of California and Nevada. Within a year he began a project to improve a road between Los Angeles and the Central Valley, taking over a cut that had been made through Newhall Pass to bypass a steep and dangerous grade. Beale enlisted a crew of Chinese workers to widen and deepen the cut to make it more passable. One of the earliest customers of the road was the Butterfield Overland Mail stagecoach company. When it was completed, Beale was awarded the rights to the cut and charged a toll for nearly twenty years.

Beale's Cut was a popular backdrop for many early western movies, a favorite of directors like D.W. Griffith and John Ford. The new highway system bypassed the cut years ago, and lies nearly forgotten except for a road marker near Caliente, California. Not content to stay on the ranch, in 1871 Beale purchased the Decatur House, across the street from the White House, in Washington, for $60,000 and proceeded to renovate it. The government rented rooms for Secretaries of State, Henry Clay, Martin Van Buren, and Judah P. Benjamin. It became the place to be for the fashionable parties of the time.

Still not ready to slow down, President Grant appointed Beale Ambassador to Austria-Hungary. European diplomats loved his stories of the Wild West and he became a popular figure in the Vienna Court. Emperor Franz Joseph I of Austria shared a love of horses that made them great friends.

Beale died at the Decatur House in 1893 leaving Mary and their three children, Mary, Emily, and Truxton. His relentless pursuit of adventure, wealth, and love of prestige provided him more adventures than anyone could ever wish for.

Boone's Road

"Through the Cumberland Gap" George Caleb Bingham, 1850-1851

In the middle 1700s, life in America was still a struggle for most settlers. The colonies were ruled by the British, and were being shaped in large part by an unstoppable flood of immigrants from Europe and other places. Although they came for many reasons, religious persecution in their homeland was responsible for tens of thousands of families looking for a new start. With the new settlers came the need for expansion. More land was needed for the growing colonies. Exploration and speculation on the unknown land to the west was a never-ending topic of conversation among the new arrivals.

In 1713, Squire Boone, emigrated with his family from England to Pennsylvania and joined William Penn and his new colony of religious dissenters. In England, Squire (his first name, not a title) Boone, belonged to the *Religious Society of Friends*, called by the disparaging term *Quakers* in England, for going against the Church of England doctrine.

Squire moved to the Oley Valley, near modern-day Reading, Pennsylvania in 1731. Working as a blacksmith and a weaver, he met and married Sarah Morgan, also a Quaker. They eventually had eleven children, the sixth of their large clan was a son, Daniel.

Daniel Boone grew up on the edge of the Pennsylvania frontier, living around the local Indians and hunting and fishing every moment he could. The steady flood of new emigrants strained the white and Indian population and the need to explore the land to the west became more important than ever.

Boone joined the British military during the French and Indian war of 1754-1763, which was a fight over the land beyond the Appalachians. After returning home, Daniel married Rebecca Bryan on August 14, 1756. By1758, the British had become involved in a conflict against the local Cherokee Indian Nation — their one-time ally against the French. Boone served with the North Carolina militia during this conflict and his expeditions took him deep into the unknown western territory. He was away from his family as long as two years at a time, and became what was called a *Long Hunter*.

Daniel Boone provided for his family working as a market hunter. Every fall, he, and a small group of men, would go on a Long Hunt, accumulating stacks of beaver, otter, and deer hides. The men followed a maze of Buffalo migration trails, called medicine trails by the local Indians. Returning in the spring, they would sell their catch to local traders.

By 1762, Boone was beginning to feel the squeeze of civilization and eventually moved to a more remote area in the Yadkin Valley, in the northeast corner of North Carolina. From there, he hunted and explored far into the Blue Ridge Mountains. Daniel had heard tales of abundant game and large, untouched acreages of fertile land from Indian traders and decided it was time to see the country for himself.

Daniel and his brother, Squire Jr., left in the fall of 1769 on a two-year long hunt into the territory of Kentucky. In December, Daniel and another member of the party were captured by a Shawnee hunting party. After losing all their goods to the Indians, they were turned loose, told they were poachers and to never return to their land. Never one to worry too much about the local Indians, Boone hunted, trapped, and explored the wild land of Kentucky for several more years.

In 1774, with the population of North and South Carolina growing and general unrest among all the colonies over the British rule, Judge Richard Henderson, of North Carolina, formed a new land speculation group called the Transylvania Company. They hoped to buy land on the west side of the Appalachians from the Cherokees and build the first British colony in Kentucky.

The judge hired Daniel Boone to blaze a trail into Kentucky to provide a route for the expected rush of new settlers. The Mountains were an imposing barrier for most settlers of the time and Boone's experience in the area would prove to be invaluable. In March of 1775, Boone, with a work party of thirty-five workers, started a road from Kingsport, Tennessee that would eventually end at the new fort of Boonesborough, Kentucky. Boonesborough would become the first chartered community in Kentucky and the first English speaking town west of the Appalachians.

The road would become known as the *Wilderness Road*, or *Boone's Road*. It was a massive undertaking for a small party of men to take on. The road started out on what was little more than a muddy footpath or game trail. Improving the trail just enough to get a horse and rider through, it would eventually be widened enough to be called a *trace* (road), allowing a full-sized wagon to pass through.

Boone and his men felled trees, moved boulders, and straightened the crooked trails, eventually pushing through the Cumberland Gap, on the border of Tennessee and Kentucky. From there, they blazed the eastern spur of the road to Fort Boonesborough.

The work of building the road produced a never-ending plague of Indian attacks from tribes such as the Cherokee, Iroquois, and the Chickamauga. Hundreds of hunters, explorers, and settlers were killed by the attacks. Log blockhouses were often built at regular intervals to provide some security for travelers.

As soon as the crude road was completed, a flood of emigrants began pushing west looking for new land. Boone became a guide, leading thousands of people to Kentucky's new frontier. Often, whole churches or communities decided to move west, and with them came their livestock. Over the next few years, tens of thousands of cattle, hogs, and sheep were moved from the colonies to central Kentucky by way of Boone's road. A new agriculture industry was soon established thanks to the new route.

By the 1870s, hundreds of thousands of Irish, Scots, and Germans fleeing starvation and intolerance, came to America for a new start. The 'empty' land was filling up fast with people looking for a better life. Daniel Boone, a giant figure in early American history, lived a life most of us couldn't even dream about today. After spending most of his life exploring and fighting in countless other battles against the Indians, the French, and the British in the Revolutionary War, he was elected and served three terms in the Virginia General Assembly.

After the revolution, he tried many different businesses over the next few years. In 1799, he moved to Spanish Louisiana (eventually Missouri), where he served as a judge and military commander. Daniel Boone's last years were spent in Missouri, hunting and exploring — always looking west for a new adventure.

Splendid Behavior

Bose Ikard in later life

As western heros go, a black man, born a slave in Mississippi with the unlikely name of *Bose Ikard* doesn't exactly leap from the pages of western lore — and that's a shame. The truth of the American West is that it was settled by possibly the largest, most diverse group of people ever to come together. Americans and immigrants of every color, ethnicity, gender, and religion put their blood, sweat, and lives into shaping the West we know today.

After the Civil War, the reconstruction and increasingly crowded eastern United States sparked a rush of people to find a place for themselves in the wide-open country of the West. In the last half of the nineteenth century, America basically threw open her doors and said "come fill up this empty land" — nearly five-hundred million acres of it. Of course, as we all know, the land wasn't exactly empty, but that's a story for another time.

Bose Ikard was one of those mostly unheralded men that helped make the West what is today. Born a slave to plantation owner Dr. Milton Ikard, in Noxubee County, Mississippi, his exact birth date is unknown but thought to be somewhere around 1843. Milton was also thought to be Bose's father, and his mother was another slave named King.

In 1852, Dr. Ikard sold his business in Mississippi and moved his wife Isabella and five legitimate children, including son William Susan (called Sude), to Parker County, Texas, about nine miles west of Weatherford, to start a cattle ranch. He also brought Bose, his mother King, and another mulatto woman slave.

At the time, Parker County was considered to be one of the most dangerous places in the West. Starting a ranch from the ground up in a part of the world where Kiowas and Comanches still raided on a regular basis, proved to be a challenging proposition for all the cattleman in the area. The Ikard family fought several battles with the Indian raiders and fortunately lost no family members in these encounters.

After a rocky start, Doctor Ikard, with help from his now strapping young slave, Bose, the ranch began to flourish. During the Civil War, William Ikard joined the Texas Calvary, with the Texas State Troopers of the Confederate States of America serving in Texas, mostly fighting Indians in the northern part of the state.

Before the war ended, the Thirteenth Amendment abolishing slavery had been signed, and Dr. Ikard decided to give his slaves their emancipation and his last name.

They continued to stay with the doctor for several years after the war. Bose, now tall and lean and looking for adventure, found his passion and his future in the cowboy life and living in the outdoors.

By fate or coincidence, two of the Ikard neighbors were Charles Goodnight and Oliver Loving. Having worked as a cowboy for years, they knew him to be a fine hand, and he even knew how to cook — a popular skill to have on the trail. Based on a letter of recommendation by the doctor, Goodnight hired Bose to work with him when he and Loving formed their first cattle drive West. They planned on taking a herd across the fearsome Staked Plains, through Horsehead Crossing to Fort Sumner, New Mexico Territory. Bose joined 18 wild, ragtag cowboys for the long drive. Many of the cowboys had served during the war, some from both sides the line. They moved out in the spring of 1866 with 2,000 head of longhorn cattle.

After weeks of heat, cold, storms, stampedes, and Indian skirmishes, they reached the fort, only to find out the government buyer would pay eight cents a pound for the 800 steers in the herd but refused the rest of the cattle. Loving decided to continue north to Denver to sell the remaining cattle. On Raton Pass, Loving had his first meeting with the famous "Uncle Dick" Wooten, former mountain man and operator of a tollgate into Colorado. Loving paid ten cents a head to get his cattle over the pass, but not without a few choice words for the operator.

The trail from Texas through Colorado covered nearly 2,000 miles and Bose quickly rose to Goodnight's right-hand man. After the trip to Fort Sumner, Goodnight and Ikard returned to Texas with $12,000 in gold and made preparations for the next drive. Over the next few years, Bose and Goodnight worked closely together and became good friends for life.

"He was the most skillful and trustworthy man that I had. I have trusted him farther than any man. He was my banker, my detective, and everything else in Colorado, New Mexico

and the other wild country. When we carried money, I gave it to Bose, for a thief would never think of robbing him."

When Loving was riding ahead in the 1887 drive, he was critically wounded in a Comanche attack along the Pecos River, but he managed to make it to Fort Sumner. When Goodnight and Ikard arrived with the herd, they found that Loving was dying of gangrene. Goodnight assured him that that his wish to be buried in Texas would be honored. After Goodnight and Ikard delivered the herd to Colorado, they returned to Fort Sumner and exhumed Loving's body. They made the 700-mile trip back to Weatherford, Texas with Loving's body together.

Lovers of cowboy history know that the book and movie *Lonesome Dove* was based on the famous story of the 1866 Goodnight/Loving drive. The character of *Joshua Deets* was based on Bose Ikard. After his trail drive days, he bought a small farm near Weatherford, married his wife Angeline, and had six kids. In 1869 he participated in a skirmish with Quanah Parker's band of roving Comanches with his former master, Milton Ikard and William Ikard.

Bose Ikard died in Austin on January 4, 1929, and was buried in Greenwood Cemetery in Weatherford. Hearing of his old friend's death, Charles Goodnight purchased a granite stone for his grave and wrote his epitaph.

"Bose Ikard served with me for four years on the Goodnight-Loving trail, never shirked a duty or disobeyed an order, rode with me in many stampedes, participated in three engagements with Comanches, splendid behavior."

In 1979 Bose Ikard was inducted into the Texas Trail of Fame. In 1999 he was inducted into the Hall of Great Westerners, at the National Cowboy and Western Heritage Museum. In 2002, Weatherford continued to honor him by naming a school the Bose Ikard Elementary School.

Burnin' Hair

Branding on the Martin Ranch, Evanston, Wyoming © Bert Entwistle

On a cold, blustery spring day, somewhere in the American West, several cowboys poke at the flaming mesquite in the branding fire, waiting for it to get the familiar red glow they need. Poking out of the crudely built fire is a pair of branding irons, each with the unique seal of its owner. As the first cowboy rides up with a calf in tow, others pin it to the ground and one cowboy slaps on the iron — or as some may call it *'burns some hair.'*

This iconic scene has been repeated tens of thousands of times every year over the course of our history and it's a common chore done by most cowboys in their day to day work.

Branding with a hot iron is one of the oldest ways in recorded history to mark ownership of property, and considered to be the closest thing to a permanent mark you can attach. Branding, of course, is not limited to only the cowboy, the livestock industry, or the American West, but can be traced back in history in one form or another for centuries. Scenes of oxen being branded with hieroglyphs are depicted on Egyptian Tombs from as early as 2,700 BC. Use of the hot iron of ownership is virtually unchanged for nearly 5,000 years.

In the earliest days of the American cattle industry, there was so much free-range land used for grazing, that branding was used to determine who owned the cattle, which

proved to be a necessity at roundup time. Many of the earliest cattle, in what is now the American West, came up from Mexico and were turned out on this range. The cows and bulls were typically branded when turned out, but the range-born calves presented another problem. From the earliest days, it was easy to find an unbranded calf on this open range and claim it for yourself. These unmarked calves, sometimes referred to as "slicks" by the cowhands, were called "Mavericks" after a Texas land baron named Samuel Maverick, who left many of his calves unbranded, and presumably unclaimed. No telling how many herds were built from these and other "Mavericks," but today the word is part of the American lexicon, used on anything from livestock to politicians and anything or anyone not connected to the larger group.

Branding laws vary from state to state, but most still have provisions for wild cattle (and horses) and what to do about them. For example, Colorado's law #35-43-118 legally defines a Maverick as: "All neat cattle and horses found running at large in this state without a mother and upon which there is neither mark nor brand shall be deemed a maverick and shall be sold to the highest bidder for cash at such time and place and under such rules and orders as the state board of stock inspection commissioners prescribes."

Use of the hot iron as proof of ownership has a dark side to it that many people today may prefer not to think about — the branding of human beings. From the days of the ancient Greeks, Romans, and Egyptians, slaves were often marked as property by their owner. The practice continued well into the 19th century in slave owning countries around the world. In many countries, convicted prisoners were branded in a misguided effort to show the public they were criminals. Even the early day college hazing rituals contained branding in their initiation rites, a practice that is hopefully abandoned by now.

It's thought that Hernando Cortez, the explorer from Spain, brought the first branding iron into the Americas in 1541 — his personal mark was three crosses. Spanish brands were often beautiful, but the complicated designs made them difficult to produce and left blurry and unreadable marks. American ranchers chose simpler designs that were hard to alter.

Altering the brand was not unheard of, especially in the early days of the American cattle business. Many a brand was altered by the later addition of a "running" iron by unscrupulous cowboys, some of which found themselves on the wrong end of a rope for their troubles. Today, the fines for stealing livestock or altering brands are still severe, but every year, many modern day "rustlers" are convicted of tampering or theft. Most states require brands to be registered, and many states have "brand books" available that show the registered brands in their jurisdiction.

Like most government agencies, the brand inspection services they provide are run by the fees collected. In Colorado, the brand inspection of "fat cattle for slaughter" costs $.65 per head plus a $20.00 fee. The brands need to be clean and clearly visible for the inspector to see them, as the cattle are constantly in motion and it's difficult to pick out the brands as they go by. A good livestock brand only has value if the brand is registered — unregistered stock causes more confusion than a Maverick. In 2007, the Colorado

Brand Inspection Board employed 65 livestock inspectors who traveled more than 1.4 million miles and inspected 5.1 million head of livestock.

The art and skill of livestock branding is basically the same now as it was back in ancient times. Today, other methods like freeze branding are used, but a good hot iron and a wood fire is considered by many to be the best formula. Cowboys who brand a lot warn against "getting your iron red hot, a hot iron the color of ashes is about right." Today, the brand on a cow or horse is as much about his owner's character and reputation as it is ownership. Generations of families are still using the same brand that their grandfathers or great grandfathers designed a hundred years before. The brand lives on long after the designer has gone to the great roundup.

Today, long after the big free-range operations are gone, Texas claims more than 230,000 registered brands on its books. Much of the history of branding and the cowboy came north from Mexico, ushered along by the Spanish Vaquero, burnin' hair as needed while they moved the herd into what would become the American southwest. The need for branding in the livestock industry is as important as ever, and the old saying, "trust your neighbors — but brand your stock" is just as true now as it was almost 5,000 years ago.

American Bred

Ready for a Good Day's Work © Bert Entwistle

"History Was Written on the Back of a Horse..."

This is an old saying that is a fundamental truth in most of the world's history of civilization. The first horses were thought to have been domesticated around 4,500 years ago. Most of the new lands were conquered by explorers and invaders on trained riding stock, and their supplies were moved by horses bred specifically for the job.

The story of the horse's history in America is an unusual one. Most of the early evolutionary development of the horse started in North America. The native animals were a small breed of horse, evolving from browsers to grazers on the plentiful grasses of the plains. Eventually, they found a route to Asia by way of the Bering Land Bridge and began to populate the Old World.

Then, for a reason that no one has been able to adequately explain, somewhere around 8,000 to 10,000 years ago, they simply disappeared from North America. In time, as the glaciers receded, the sea swallowed up the land bridge, preventing their return. The horse, originally an American, had been exiled from its ancestral home.

It was nearly the sixteenth century before the next horse set its hooves on the soil of the new world. On Columbus's first voyage in 1492, he did not bring horses. When the King of Spain realized the potential of the new discovery, he ordered that all ships traveling to the new land under a Spanish flag must carry horses. He quickly understood the value of the horse for the exploration of the new region. The King wrote to his secretary and asked him to choose twenty fighting horses (fighting men generally rode stallions) and five dobladuras (extra horses, usually mares) for the next voyage.

When the second expedition left Spain in September of 1493, there were seventeen ships and 1,500 hundred people under Columbus's command. Others on the voyage also brought horses with them, ready to lay claim to the wealth of the new world on the back of a good horse.

For the next twenty years, Spanish explorers mapped the Caribbean, lower Florida and the Gulf Coast. In 1519, Cortez landed on the mainland and entered Mexico City with sixteen horses. The transcript of the voyage recorded Captain Cortez as having a "vicious dark chestnut" as his personal mount. They also noted that "the natives had not seen horses up to this time and thought that the horse and rider were all one animal."

By the 1520's, horses inhabited all of the Caribbean islands and were getting a solid foothold in Mexico and the coastal area. In the winter of 1521, Ponce de Leon landed in

South Florida with the goal of establishing a permanent colony. Although his effort failed, it was the first time large numbers of horses had been on the Florida mainland.

Within six months of Cortez's landing, many more Spaniards landed in Mexico with many more head of horses. The first recording of branded horses was in 1540, and the first horse race of note was on December 27, 1540 between Francisco Vasquez de Coronado and Rodrigo Maldaonado. It was written that Coronado fell from his horse and was hit in the head.

Coronado led the first major expedition of men into the American Southwest in the summer of the same year. They discovered the Hopis and the Navahos, and the natives saw horses for the first time. Although it was widely thought that the first wild horses all came from those lost by Coronado, there appears to be enough evidence to believe a good share of them came from early trading, and theft by the Indians.

In 1623, a friar, named Benavides, made the first known reference in his journal of seeing a Gila Apache war chief riding a horse. At that time, the Spanish forbade trading horses to the Indians. In James A. Michener's wonderful novel *Centennial,* a story of the settling of Americas high plains, an Arapahoe warrior named "Lame Beaver" steals their first horses from a Comanche war party near the Arkansas River in Colorado and became a legend in the tribal history. The story details how important the horse was to the native culture. Any tribe that was without horses by the mid to late eighteenth-century would soon be lost to history.

The horse allowed the plains tribes to be truly nomadic in the sense that they could now roam great distances and trade with other tribes as well as the white men moving into the open prairie country. The wild breeds of horses, eventually called Mustangs, may have originally been descended from the Spanish Barb, but herds of wild horses from the Eastern U.S. crossed the Mississippi and began to mingle with the western horses.

By the nineteenth-century, the wild herds were so thoroughly mixed, that you could find every color mix, size, shape, and conformation ever seen. Horse trading among the tribes and the settlers, and even the army, became big business and the country's economy became dependent on the horse. The horse had been proving its value in the Old World for several thousand years. Now the new world was growing and rapidly expanding its horizons thanks largely to the adaptability of "Equus ferus" (the wild horse) and "Equus ferus caballus" (the domesticated horse).

Throughout America's history, the horse was there, exploring the new land, turning the soil, pulling wagon loads of goods or delivering the mail. It has carried our soldiers into battle from the Revolution to the Civil War and proved invaluable in countless victories around the world.

The wild herds of Mustangs were left largely unregulated well into the twentieth century, when the large sections of open range were turned into grazing land for domestic animals. At the turn of the century, the wild population was estimated to be about two million animals. In 1971, Congress passed the Wild Free-Roaming Horse and Burro Act which gave protection to these animals.

Like so many well-intentioned efforts, the protection caused a problem of another sort—the horses soon multiplied to the point of overrunning the range. The new act required the removal of all excess animals to restore a proper ecological balance on the range. Today, programs like Adopt-A-Horse make it possible for people to adopt, for a fee, one or more horses.

Throughout history people like General Washington, General Custer, and President Theodore Roosevelt have made and recorded history from the backs of their horses. Unnamed American Indians like Michener's fictional Lame Beaver brought the horse to their tribes and added both oral and written versions to their own history. After thousands of years of absence, the horse had returned to North America just in time to help record its history.

8 Cows and One Stud Bull"

"Unbranded" C.M. Russell - 1897

What would it be like to get the boot out of your own country after nearly a hundred years? Where would you go and what would you do when you got there? What if you had to leave everything behind except what you could carry? This was the scenario called the *Great Expulsion* of 1755–1763.

After a hundred and six years of French control, starting in 1604, the British captured Port Royal, Acadia (Nova Scotia), and began to rule the French citizens with an iron fist. The stubborn Frenchmen refused to sign an unconditional loyalty oath to Great Briton, and open resistance became common. In 1755, Great Briton could no longer tolerate the thought of their old adversaries living on British soil, and ordered more than 11,000 people, nearly three-quarters of the population, deported from the Acadian Peninsula. Homes were burned, land confiscated, and, in some cases, families split up and people imprisoned.

After losing everything in the North Country, a few families eventually found their way to southern Louisiana. Although it was a totally unfamiliar landscape and climate, they hoped the resident French population would welcome their fellow countrymen. At the time of the French deportation, Louisiana was controlled by France. However, by the time they made their way to the delta, France had lost the territory through the Treaty of Paris, and it had been turned over to Spain.

Though still firmly in the hands of French locals, the Spanish were starting to flex their muscles. What little French currency the travelers had was not recognized, and after long and arduous trips by ship and overland, the new French immigrants landed in Louisiana with little or nothing to start their new life.

The "Acadians" as they were called, started arriving in 1764, and by 1768, over a thousand refugees had found their way to their "New Acadia." After landing in New Orleans, Commissioner Nicholas Foucault was faced with the problem of where to put so many new people. Records show that he was ordered to give them nothing but the bare essentials, but Foucault knew he had to do what he could to help them. He spent 15,500 Liveres of government money on construction material, tools, guns, and food for the new arrivals.

The Territorial Governor, Charles Philippe Aubry, knew he had to find a permanent location for them to settle. Raw land was still plentiful and he first thought of the east bank of the Mississippi River. The east bank was still a thick, rough tangle of hardwoods and flooded every year. He finally decided on what was called the Attakapas region, named after one of the local Native American tribes. The Attakapus Indians (locally called the man-eaters), had recently been moved out of the country. The region was mostly open prairie, and perfect for immediate farming.

A military engineer, Louis Andry, was appointed to lead the Acadians to their new homes and lay out a village and the land grants. They were provided with hardtack, flour, rice, and salted meat for six months, as well as plenty of seed for corn and rice. Andry and the new settlers laid out the grants for each family, the larger the family, the larger the piece of ground. The only thing missing from this new life was livestock.

The live cattle industry had started out on the Louisiana prairie well before the Acadians arrived. The first French brand was recorded in the local French brand book in 1739. When the new Cajuns (from the word Acadian) started farming, they wanted to buy hogs and cattle, but had little or no resources to do it. The Cajuns again benefited from the generosity of a stranger named Captain Antoine Bernard d'Hauterive, a wealthy local French colonial official. He agreed to loan the Cajun families "eight cows, and one stud bull for a period of six years, afterwards the settlers would return nine cattle and half the offspring produced."

Most of these Cajuns had been successful farmers in Acadia, and raising livestock was routine practice to them. In Louisiana, with its warmer climate, the farmers began to thrive. By the 1770's, they averaged at least 22 cattle per grant and owned at least six horses. By the 19th century, each "vacher" (ranch), had at least 100 head. The larger vachers began to bring their cattle to New Orleans for sale in the local markets. Two locals, Amant and Pierre Broussard started to drive their cattle to the city along the Old Spanish Trail, generally taking 100-150 cattle at a time.

These early Cajuns ran mostly the descendants of the free ranging Spanish cattle that roamed much of Louisiana and East Texas. Generally black, and blessed with some of the longest horns they had ever seen, these lean, rangy cattle came with an attitude, but were soon domesticated. To keep them on their property, they built miles of fences called

"Pieux" fences, an ancient type of stout, split rail construction to keep them in. In 1761, the first Cajun cattle brand was officially entered in the French brand book. Branding was considered a necessity for the vachers considering the difficulty of rounding up and holding all the unbranded cattle in the area.

The Canadian refugees blended in with the early French, German, and Spanish settlers and learned much of the cattle business from the Spanish Vaqueros and local natives like the Avoyelles tribe. Horses, now more plentiful than in the north, proved to be a huge improvement to the Cajuns. They were now able to locate and work their cattle faster and more efficiently than ever before, as well as use them for draft animals. Cajun cattle drives continued to provide beef to New Orleans as well as any surplus corn or rice.

Today, Cajun country is generally considered to be most of southern Louisiana's rural prairie and wetlands. As more Canadians moved in, they mastered nearly everything to do with agriculture, including growing rice and fishing. Things like alligators and rattlesnakes became not only part of their life, but part of their culture. Today, the Cajuns are as well known for their food and their music (called "Zydeco"), and their mix of languages and accents as they are for their beef. What was done by a group of fresh immigrants with "eight cows and one stud bull" was to create a great new culture called "Cajun"

Canyons, Cows and Six-Guns

"Smokin' Them Out" C.M. Russell - 1906

Imagine what your favorite cowboy movie or novel would be like if it were set in the cornfields of Illinois, the bayous of Louisiana, or on the beaches of Florida…What if the posse was chasing the bad guys through the streets of New York City…? Something would be missing, something that is common to most western novels, movies, and TV shows. That something would be the canyons, mountains, and deserts of the American West.

The area of the west known as the Colorado Plateau could be the setting for nearly all the western stories ever produced. Roughly centered on the Four Corners area, it takes up large chunks of Utah, Colorado, Arizona, and New Mexico and is one of the most scenic areas in the world. Within its boundaries is the greatest concentration of national parks and monuments in the United States. From Dinosaur National Monument in the north to Grand Canyon National Park in the south, and Mesa Verde National Park to the east, the landscape is a fantastic maze of deserts, forests, rivers, and canyons full of spectacular natural arches.

The Colorado portion, called the Uncompagre Plateau, from the Colorado National Monument south was occupied by the Ute Indians when the first white explorers, two Franciscan Friars from Spain named Dominguez and Escalante, traveled through in 1776. In 1875 the Hayden survey party named several local streams and canyons after the Friars.

By the time Colorado achieved statehood in 1876, the population was demanding the removal of the Utes of western Colorado, so that any land that could be "mined, farmed or ranched" would be opened for development. In 1880, the Northern Utes were forcibly removed to the Uintah Valley Reservation and the western slope of Colorado was thrown open to settlers and entrepreneurs of all kinds.

In a short time, the area was covered with mining claims, farms, and livestock ranches. The railroad arrived in the area in the late 1880s and by 1900, tens of thousands of cattle were grazing every corner of the plateau. The railroads also brought thousands of sheep — their owners determined to get a fair share of the newly opened country. They pushed their herds onto the same public lands the cattlemen used, and the stage was soon set for Colorado's own cattle and sheep wars.

By the early years of the 19th century, the herds had penetrated every possible spot that any forage could be found and the ecology of the area began to change. The ranchers were accustomed to grazing their cattle when, where, and however they wanted, and it was up to them when they moved their stock.

The sheep men wanted the same rights and often put their stock on the range before the cattle or after the cattle had left, to graze it even more. The same feud erupted on the plateau that had happened in the west many times before, and in many cases, things between the ranchers turned ugly.

The range became so badly degraded from a generation of overgrazing, that the sheep ranchers and cattle ranchers began to point fingers at each other as the cause of their problem. The forest ecology began to change, and the periodic fires that nourished and helped regenerate the forest stopped for lack of fuel on the ground.

Passions and tempers between the cattlemen and the sheep men rose to a fever pitch and soon livestock on both sides began to suffer. Sheep were shot, and ran off cliffs and left to die. Soon, instead of pointing their fingers at each other, they started pointing guns.

June 9, 1917 was said to be a warm, sunny, blue sky day so typical of Colorado. Escalante Canyon, a spectacular ancient place full of fractured rimrock and rock art mixed in with Pinon and Cedar trees was just warming up. Deputy U.S. Marshall Cash Sampson, though a small man, was still lean and fit at 46 years-old, and rode alone through the desolate canyon with the intention of checking on a few of his cows up on a remote mesa. On the way, he was flagged down by another local rancher, Kelso Musser, and, as was the custom of the day, was invited in for a bite to eat. Sampson thanked Musser and dismounted, ready for some hot biscuits and a little conversation. Within minutes, another cattle rancher, Ben Lowe, rode in with two of his young sons alongside of him and joined the breakfast conversation.

At the table the small talk was pleasant if not a little restrained. Sherriff Sampson, a cowman his whole life, had been charged with enforcing the law, and nightriders had been killing sheep and harassing local sheep ranchers. He knew that he would have to step in eventually, and here, across the table from him was Ben Lowe, generally accepted to be a leading player in the nightrider's dirty work.

Ben Lowe, onetime friend of Cash Sampson, saw the little marshal as a traitor to the cattlemen everywhere and knew that Sampson was scheduled to speak against him in an upcoming grand jury investigation. Lowe had trouble sitting silently and decided it would be for the best if he just moved on. He and his boys mounted up quickly and headed out.

Within minutes, Cash Sampson was on the trail behind the trio. Catching up quickly, Sampson rode up behind them. When Lowe saw the marshal, he told his sons to ride on ahead, he would stay to talk to Sampson and catch up in a few minutes. The boys were hardly around the first bend when they heard the gunshots. They raced back down the road to find Cash Sampson with a bullet hole in his head, sprawled out dead in the road, horse grazing nearby. Ben Lowe was slumped awkwardly against a Cedar tree, near death with blood flowing from his chest. As the boys ran to their father, he fired one more shot into Sampson's body before he slid the rest of the way to the ground and died.

When local ranchers got to the scene, Lowe's 45 revolver laid along side of him with three rounds expended, and Sampson's 32-20, still firmly in his grip, showed only one fired round. Although the locals took sides and offered many theories as to exactly what happened, no one will really know just what was said that day to turn a beautiful, lonely canyon into a scene from an old western novel set deep in the heart of the American West.

Oklahoma City's First Citizen

U. S. Marshals Bill Tilghman and Charles Colcord in 1893

Charles Colcord's earliest memories of his father was that of a striking looking man in a confederate gray officer's uniform. Born August 18, 1859, in Bourbon County, Kentucky, to William Rogers Colcord, a colonel in the Confederate States Army, and Maria Elizabeth Clay, also of Kentucky. At the start of the Civil War, William moved his family from Kentucky to Georgia. After the war, wanting to get away from the worst of the turmoil of southern reconstruction, he sold his share of the family farm in Kentucky to his brother and bought a sugar plantation near New Orleans.

Young Charley, always looking for adventure, spent his time exploring the local swamps and eventually contracted malaria. To help him recover faster, his father sent him to Banquete, Texas, to live with a rancher friend named Charles Sanders to recover and learn the cattle and horse business. After his father opened a horse ranch near Corpus Christi, he left Sanders to work for his father as a cowboy.

In 1875, at age sixteen, his father sent him on his first cattle drive from Corpus Christi to Baxter Springs in the southeast corner of Kansas, a trip of more than 800 miles. On the drive through Kansas, Charley saw a need for more horses and encouraged his father to trail a herd north to sell.

In 1876 William partnered up with another rancher, Hines Clark, and trailed 1200 mares north to the Cherokee Outlet. Once they were delivered, Charley decided to stay in Medicine Lodge, Kansas in Comanche County, just above the Oklahoma border and in the heart of Comanche Indian territory.

By the fall of 1877, the rest of the family were all living in the area. His father then formed the *Jug Cattle Company* with several other area ranchers and Charley was made the range boss for the operation. Eventually, the Jug brand would become one of the most famous in Oklahoma and Kansas.

The Jug Cattle Company joined fourteen other brands and formed the famous Comanche Pool, one of the first large corporate outfits in the southwest. The pool started with 26,000 head, controlling 11,000 acres, and was sending 20,000 head to market every year.

Henry Brown and John Middleton, fresh from Billy the Kid's outlaw gang, worked at Colcord's cow camp in 1879. Middleton married Colcord's fifteen-year old sister, Maria *Birdie* Colcord. Fortunately for the family, the marriage was short lived.

By the fall of 1882 the combined membership of the pool eventually gained government leases for ten years and grew to over 84,000 head controlling more than 3,000,000 acres. For this they paid an average annual price of two cents an acre. With cattle prices high, the pool members continued to grow their herds until Washington began to hear rumors of corruption in the operation. President Cleveland immediately voided all the leases.

The government ordered more than 200,000 cattle removed from the government land within forty days causing serious overstocking issues in adjacent states. A drought in 1885 was followed by a particularly bad winter causing an 85% loss in Comanche Pool assets. The majority of the pool members were forced out of business and left the country. The Jug Cattle Company hung on until the bitter end and went out of business like the others.

In 1889 the Oklahoma Land Run was announced and Charley decided to take his chances on the free land. The most popular version of the legend has it that he made his run on April 23, 1889, and when he found what he wanted, he traded his wagon and team (a $66 investment) for a shack and the lot. His lot is said to be Lot Number 1, Block Number 1, in the newly minted town of Oklahoma City.

The new town was a wild, dusty, collection of tents and hastily built log and clapboard buildings. As one of the most well known people in the area, he was appointed chief of police and then became the city's first sheriff. He became so well respected as a lawman that President Grover Cleveland appointed him Deputy U.S. Marshal serving with famous lawman Bill Tilghman. As a marshal, he fought against outlaws like *Bill Doolin, Little Dick West,* and *Tulsa Jack Blake.* He hunted down and captured five members of the *Dalton Gang* and personally supervised their hanging.

In 1893 he participated in the Cherokee Strip Land Run, and ended up building a home there. When the dust settled, Perry, Oklahoma had a population of 15,000 in just over six hours. Charley got the call and was appointed marshal to Oklahoma's newest town.

In 1898 he moved back to Oklahoma City, now a booming cowtown and decided to try his hand at land development and real estate. He started the Colcord Investment Company and the Colcord Park Corporation. He also started the Commercial National Bank of Oklahoma City, served as the vice president of State National Bank, and

President of the Oklahoma City Building and Loan Association. Colcord's business sense led him into the oil business, and in 1901 just north of Red Fork, he and partner Robert Galbreath, formed the Red Fork Oil and Gas Company, bringing in many strong wells in the area.

In 1905, while hunting with partners Robert Galbreath and Frank Chesly, his two wolfhounds took chase after a wolf and disappeared. Searching for the lost dogs, they found them on a farm owned by a Creek Indian named Ida Glenn. During the hunt for the dogs, they passed a spot that had an oily substance seeping from some rocks. They made a deal with the owner and on November 22, 1905, the first well they drilled gushed. They named it the *Glenpool* oil field. It became one of the richest oilfields in the world pumping millions of barrels making Tulsa the *Oil Capitol of the World*. Robert Galbreath became known as the *Oil King of the Southwest*.

In 1912 Charles Colcord built the Colcord Building in Oklahoma City for $750,000. Today it's known as the Colcord Hotel and is on the national Register of Historic Places. In the 1920's, he built a new ranch in Delaware County and nearby a new town was taking form, they named it Colcord. He became one of the most loved characters in the state of Oklahoma history and collected many more honors over the years. He was inducted into the Oklahoma Hall of Fame in 1929.

Colcord had one last reminder of his early lawman days when his good friend Charles Urschel was kidnapped in 1933 by infamous gangster Machine Gun Kelly and held for ransom. Colcord pulled together a group of wealthy local businessmen and put together a large reward for his capture. Kelly was captured and Urschel was released unharmed.

Charles Francis Colcord married Harriet Scoresby on February 9, 1885 and had seven children. Charles passed away December 10, 1934. His body lay in state in the rotunda of the Oklahoma Historical Society building with thousands of Oklahoma residents paying their respects to the First Citizen of Oklahoma City.

Confederate Cow Cavalry

"The Beefsteak Battle" from an 1874 engraving in *Harper's Weekly*

The American Civil War, the brutal and horrific conflict that cost more than 620,000 Americans their life, lasted four years and became the world's first serious mechanized conflict. Trains, telegraphs, ships, and vastly improved weapons were now a standard part of warfare. They all worked in concert to locate the enemy and move quickly to engage them in a fight.

With several million soldiers in combat throughout the conflict, the ever-increasing speed of war began to expose another problem. The logistics of keeping your army supplied and fed was always a tough job, but for the Confederates, it became especially difficult.

As the war drug on, the North used every means possible to stop supplies from reaching the southern armies. The northern ships blockaded every seaport and river, and their troops destroyed the railroads and burned the fields. The Yankees were better equipped from the start, and had many more reserves to draw from. As the union army began to cut off their supplies, the results were dramatic. The Confederate Army knew they would slowly starve if they couldn't find a steady source of food.

As noted by historians and journalists of the period, when the Confederate prisoners came into the camps, many were emaciated and had obvious health problems, not related to wounds that they had received in battle. Their uniforms were generally loose and baggy and most were poorly made and seldom patched. Most of them asked for food and then tore into it ravenously when they got it. The longer the war lasted, the more the north pushed to cut off all the southern supply routes and bring a quicker end to the war.

In the 1860s, wild cattle were common in Florida. Descendants of the original strains of cattle brought to Florida by the Spanish in the 1500s the wild cattle had regularly been caught and used by the locals for their own needs. These were considered to be scrub cattle, scrawny and small, and extremely wild and unpredictable. As the confederate need for fresh meat grew, Florida officials decided to step up and do their part to help the cause. These cattle could be collected and sent north to help the confederate cause.

Special units of militia were formed, using men that had experience in the livestock business. They named them *The Cow Cavalry, 1st Battalion Florida, Special Cavalry.* The battalion consisted of nine companies, each with a hundred men. These troops did everything from gathering the wild cattle from the frontier areas of Florida, to driving the cattle north to the confederate armies that needed it most.

Until now, the Florida cattlemen (often called cowhunters), had been exempt from the draft as they were already selling at least part of their herds to the south. On February 17, 1864, the confederate congress changed its rules of conscription, eliminating the draft exemptions of any cattlemen providing food to the confederate cause. The congress then sent teams of conscription agents to Florida to locate and sign them up — or to arrest them. For the first time, the cattle hunters were forced to declare their loyalties. A few choose the Union Army, but most went ahead and pledged their allegiance to the confederacy.

These wild scrub cattle and the cowmen who knew how to catch them were the one thing keeping the confederates from starving near the end of the war. These cattlemen had become known as *Crackers*, for the whips they carried to push the cattle out of the thick scrub. The Crackers found themselves part of the newly formed Cow Cavalry, rounding up and driving cattle from Florida to Savannah and Charleston, from May through to the first frost of fall in northern Florida. They also drove hogs and sheep north, and provided fish, fruit, and salt for the southerners.

Throughout the fall and winter, they protected the herds from northern raiders and local cattle thieves, and did all the general work required of any regular soldier. They went on missions to help locals that suffered losses at the hands of Union soldiers, and joined in many battles against the Union troops.

Skirmishes and running battles with the northern raiders were frequent and caused many casualties and deaths during the *Cow Cavalry* operations. In April of 1864, John T. Lesley, captain of B Company, engaged the northern troops at *Bowlegs Creek,* in Polk County. This cost the company their first losses of the war, losing two men. In July of the same year, B Company attacked a force of 800 Yankee soldiers caught burning, looting, and destroying the local property along their march route. The Yankees had landed their troops at the coastal town of Bayport, and began searching out any confederates in the area. Lesley and his Cow Cavalry company, were instrumental in setting up a trap for the marauders and managed to turn them back toward their boats in a hurry.

As the outcome of the war became obvious, Lesley was called on to do one last thing for the confederate cause. In May of 1865, a quiet, bearded man introduced himself to Lesley at the captain's house in Brooksville. After a few minutes of conversation, the man identified himself as Judah P. Benjamin, former Secretary of State for the confederacy. Benjamin feared for his life and appealed to Lesley for help to escape the United States.

Lesley arranged for a boat to take him to Ellenton, Florida, where he stayed at the Gamble Mansion. From there, secret passage was arranged from Sarasota to the Bahamas and then on to England where he lived out the rest of his life.

The Cow Cavalry played an important role in the history of the Confederate Army, and the history of Florida, especially in the last year of the war. After the war, Florida went on to be one of the greatest cattle producing states in the union. In recent times, Florida had the bragging rights to having more cattle than any state in the union. Its temperate weather, and plentiful rainfall producing more than enough feed for tens of thousands of cattle.

The history of the Cow Cavalry helped shape the legacy of Florida as a powerhouse in the live cattle industry, and became the home of the original Cracker Cowboys.

Feed Lots & Cowboy Brogue

Chicago Union Stockyards about 1880

Did you ever have a moment of inspirational clarity (an epiphany if you will), that caused you to doubt the accepted version of something? At some point in your life, did you begin to realize that maybe George Washington didn't really confess to cutting down old pop's favorite cherry tree? Some say history is recorded by the conquerors of society — and no doubt a lot of it is. However, in the early days of this country a lot of the population could not read and write, and the history that didn't get written down was passed on orally. As most of us know from grade school, that doesn't work very well. He who talks the most and the loudest with his version is often the one most often remembered.

The story of the live cattle industry in this country is full of wonderfully romanticized stories of wild cattle and Texas trail drives. We all know that after the Civil War, the Longhorns ran free and the great Texas range cattle business got its start there. These old stories have been told and retold so often and so loudly that many people believe that this was the beginning of the trail drives in the United States. Those folks would be wrong. Just a case of hearing the old stories over and over for the last hundred years and not looking any farther back.

More than two hundred years before Texas, or any other place in the west was even considered part of this country, a man by the name of William Pynchon, his slave, Peter Swank, and his son, John, would lay the first foundations for the first trail drives in the new world. They also opened the first known feedlot business and the first meat packing company.

Pynchon was one of the earliest colonists, moved by the British intolerance to his Puritan religion, to immigrate to the colonies and become a businessman in the new world. He became very successful as a fur trader, farmer, judge, and founding father of Springfield, Massachusetts.

William Pynchon, and his son, John, were deeply invested in the local livestock industry when Peter Swank came to him as a new slave around 1650. William had wanted to sell cattle to Boston for years, but knew the hundred-mile trip would be too hard on the cattle. In the spring, often called *the starving time* in those days, the cattle were gathered and brought back to local farms. They had been left to the range on their own all winter and were little more than hooves, hides, and bones. It took them months to fatten up on local grasses, and no market wanted the scrawny cattle. In many cases, the early settlers weren't much better off themselves.

Peter Swank, fresh from the slave trading centers of the Caribbean islands, had tended the cattle for the shipping companies and the buccaneer's camps. In time, he explained that there were no wild grazing cattle on the islands and that they were kept penned up and fed until they were needed. In the colonies this was unheard of, but

Pynchon understood the merit of Swanks system and started to move his cattle operation in that direction.

William was not only a successful businessman, but a devout Puritan, and in time he became very frustrated with the direction the religion was taking. In 1652 he wrote a book called *The Meritorious Price of our Redemption,* about the decay in Puritan morality and was soon publicly denounced by the religion as *ungodly.* Its sale was banned, and the book was burned in Massachusetts. Pynchon became the first victim of censorship in the new world. He felt so shamed by the reaction, he soon turned his business over to John and sailed back to England and out of the colonies forever.

John had been preparing for Peter Swanks *feedlot* system, and for the next winter or two, while the rest of the farmers turned their cattle out, the Pynchon cattle were fattening up all winter in their stalls and pens. Just as Swank had taught, his cows came through the starving time fat and healthy. Pynchon immediately started to initiate the next part of the plan — hire some men to move the cows from Springfield to Boston.

Cowboys wanted: red hair, freckles and a strong Irish Brogue desirable .

When Pynchon went looking for people to work the herd, he turned to the large population of Irish immigrants in the area. Already known for their cattle and horsemanship skills, he had a ready labor force to work the cattle. Ireland had been a center for beef and milk cattle for hundreds of years. The smallish Irish horse was already established as a cattleman's breed and the term *cowboy* was firmly entrenched in Irish history and music. This horse was considered to have a gait "as comfortable as a rocking chair on the hob." The flat stone floor in front of the fireplaces of the day was called the *hob*, and considered the most comfortable place in the house. The term *hobbyhorse* came from this old saying.

The Irish men that Pynchon recruited were nothing if not a diverse lot. Some came to escape the tyranny of the most recent English conqueror, Oliver Cromwell, some were escaped slaves and indentured servants and it's likely that a few were looking for adventure. When the spring rolled around, Pynchon and his new *cowboys* were ready to roll.

After recruiting Irishman, John Daley, as head cowboy, and, John Stewart, as blacksmith, the men gathered them into Pynchon's pens and branded them. When first they gathered them up they were wild, ornery, and mean. In the spring, after a winter of being well-fed and tended, they had become fat and gentle.

On a cold spring morning in 1655, John Pynchon and his bunch of newly minted Irish cowboys, headed out for the first-ever drive of fat cattle inside the modern boundary of the United States. As they opened the pens and pushed the cows on the trail to Boston, they rode right out of Pynchon's farm and into the history books. Several weeks later, after selling his cattle in Boston for a good profit, he made the decision to do this every year. You could rightly call John Pynchon America's first cattle baron. All this was more than 121 years before the American Revolution. Sadly, there is no written record of the weeks on the trail of this first drive except for Pynchon's hand-written notes done in some type of personal shorthand that has never been deciphered. A personal written account of America's first trail drive?

What an incredible read that would be . . .

Waiting for a Chinook

Charlie's Answer to an Investor in the East About How Bad the Storm Was

By the early 1880s, the great open range cattle industry in Wyoming had reached its peak and was beginning to show signs of weakness. The fabled cattle barons of the day were having trouble keeping the flood of new settlers from claiming their piece of the great western land giveaway. The ranchers considered it to be their range by virtue of them running their stock on it in the past. The way they saw it, nothing was going to stop them from their way of doing business. It was often war on the open range, with settlers being threatened off their land, and groups of men were hired to "convince" them to leave their legal claims. When that failed, many were burned out and even murdered.

The inherent weakness of the open range system and the refusal of the ranchers to accept the need for change spelled the beginning of the end for the old ways. Falling beef prices, combined with serious overgrazing of the range by the cattle and the new sheep industry, left little feed for either. The summer of 1886 was brutally hot and dry, and the badly damaged range weakened the cattle even more.

The new citizens of Wyoming continued to stake their claims in defiance of the cattle barons, and when they came, they brought the tools for working the land with them. They also brought the natural enemy of the open range cattleman — barbed wire. Most of the new transplants were more interested in farming than cows, and the wire was intended to keep the hungry cattle out of their fresh fields. The winter of 1886-1887 was about to show all of them who the boss really was.

Charley Russell

Saturday, November 13, 1886, marked the first day of what could be called Wyoming's "Perfect Storm" — or, the day the open range cattle business died. After a crisp, mostly clear day, snow clouds moved in and began to darken the prairie. The snow began to fall, and it came down almost continuously for a month. Heavy wet snow at first, and then the temperature began to fall, and the snow got deeper, leaving the cattle unable to move around, or reach what little feed was left.

By January of 1887, the temperature warmed just long enough to freeze the top layer of snow. Then overnight, the storm plunged the open, wind-swept grassland into sub-arctic temperatures. Monstrous blizzards of ice and snow scoured every inch of the land. The ability of the cattle to withstand the harsh climate of Wyoming's winter was one of the most prized traits in the range bred cattle, but this was unlike any storm anyone had ever seen.

The huge cattle operations of the day gave little consideration to their wild, roaming stock scattered over hundreds of square miles, and often never saw them until they were gathered for branding or market. The free-ranging herds looked for shelter anywhere they could, finding every gully, tree, rock, and low spot in the ground. Temperatures as low as minus 49 degrees were recorded that January. When one animal found a small piece of shelter, hundreds more pushed their way in, and soon the snow leveled out across the ground and they disappeared forever — or at least until the last of the snow melted.

Cattle died by the tens-of-thousands. Stories of frozen cattle still standing upright were common, and it wasn't until late spring that the true extent of the devastation was realized. The bloated bodies littered the landscape as far as the cowboys could see. Every stream, river, and water tank was filled with piles of corpses. As the ice went out, the bodies floated down the rivers and, in some cases, created dams that had to be cleared out.

For all practical purposes, anything that was outside the confines of the ranch was dead. Also dead was the open range cattle business. The great operations of the day had lost nearly everything, and the system that had made them rich was gone forever. Wyoming lost many of its new settlers and more than a few of its cattle ranchers that spring. Bankrupt and broken, most returned to their roots in the east.

The locals called it the "Great Die-Up." Many well heeled and well known people of the day, like Theodore Roosevelt, and wealthy investors from as far away as the east coast and even royalty from England, were in the cattle business. No one was spared.

In 1886, a young itinerant cowboy named Charley Russell, was working at the OH-Ranch in the Judith Basin of Montana when the storm hit. At 22-years old, he had worked hard as a cowboy, but deep down inside he considered himself an artist. By spring, the results of the blizzard were already becoming legend. The OH had suffered the storm nearly as bad as Wyoming and their losses were devastating.

One night, sitting in the bunkhouse of the OH with the cowboys, was the owner, Jesse Phelps. He had recently received a letter from fellow cattleman, Louie Kaufman, from Helena. Kaufman was inquiring about how the OH cattle were doing. Phelps told Charlie that he had to write a letter to Kaufman and tell him how bad things were.

After a little more conversation, Charlie came up with the idea of sending him a sketch to go with the letter. Finding a piece of scrap cardboard, he did a small watercolor he titled *Waiting for a Chinook* (warm wind). He subtitled it *The Last of 5000*. The small picture depicted a starving, emaciated steer, near death, surrounded by hungry wolves. After one look at the picture, Kaufman decided he didn't need much of a letter, the picture said it better than he could ever write it. That small picture became famous all over the west.

Eventually, Charles Marion Russell went on to be one of the most famous and loved American artist of all times. He had lived and worked with the cowboys and Indians of the time, painting from memory the details of the fading west. Finally getting married and settling down in a home in Helena, he continued to paint and sculpt up to his death in 1926.

Years later, Russell repainted the drawing that got him so much attention in a larger version, because of its popularity. The original postcard sized work hangs in the Buffalo Bill Museum in Cody, Wyoming. As the real west began to fade, Charley Russell was there to get it all down, storms and all. How lucky we were to have Charley remember it for all of us.

Long Island Aged Cattle

Long Island, New York. Home of the United States' First Cattle Ranch

On a clear day, you just may be able to see it from the top of the Empire State building, way out there on the tip of Long Island, nearly surrounded by the Atlantic Ocean and Block Island Sound. If you can see the Montauk Lighthouse, you are looking at a genuine piece of American history commissioned in 1792 by President George Washington. The lighthouse itself is surrounded on three sides by thick rolling grasslands, land that was being used for summer livestock pasture more than a hundred years before the lighthouse was built. Montauk, on the far end of Long Island, is considered by many to be the site of the oldest cattle ranch in the United States and the home of the original cowboys.

Although these claims can hardly be proved to be rock solid, it's for sure that Long Island did have a great natural summer pasture and beef on the hoof was an early form of wealth for the settlers, often referred to as "Colonial Currency" by the locals. When the first explorers and settlers came to Long Island in the early 1600s, they found Native Americans of the Algonquian band led by Montauk Sachem in control of the island. From the beginning, the local Indians made many deals to the settlers to lease their land using cattle as payment. Even the new government would accept cattle and other livestock as payment for new land patents.

From the mid-seventeenth through to the late nineteenth century the Long Island ranchers turned out their stock on May 1st, historically called "Cattle Day," driving them as much as 70 miles to the thick green pastures on the tip of Long Island. The custom of the day was to give the cattle a registered earmark for easy identification at roundup. In the early days, after making their deals with the Montauk Indians, they turned their cattle, sheep, and horses out on the long unfenced peninsula for the summer. The first of November marked roundup time and Thanksgiving in Montauk was held the first Thursday after the successful completion of the drive. Eventually, the local Indians began to sell some of their land to private owners.

From 1660 to just after the turn of the twentieth century, the system worked pretty much unchanged. In this period anywhere from 2000-6000 cattle, horses, and sheep roamed the grassy rolling hills of Long Island mostly unattended for the summer months. Throughout the 1700s only three houses existed in the pasture country, they were called simply First House, Second House, and Third House. Each house was spaced three miles apart and housed the summer keepers (early cowboys). Second and Third House are still standing and both house museums today.

By the time of the American Revolution thousands of cattle, sheep, and horses were on Long Island. On August 22nd, 1776 the Battle of Long Island was raging, and the islands stock was rounded up and driven east to Montauk to try and keep them from the British. It was an early cattle drive that ended badly with the locals losing both the battle and the livestock.

The history of the island over the years has been colorful to say the least. Stories abound, ranging from Captain Kidd burying two chests of treasure in Money Pond (nothing found to date), to Prohibition era Rum Runners burying their illegal brew in the dunes. During the 13-years that alcohol was against the law in the United States, the Montauk area was a well-known area for smugglers, and locals say that even now some of their liquid loot turns up after a big storm. In 1898, Teddy Roosevelt and his "Rough Riders" were quarantined in Montauk, recuperating from the war in Cuba. Teddy himself stayed in Third House at Indian Field, now in the Montauk County Park system. During World War II, there were German U-boats sighted off the beaches.

The first private ownership of the land was in 1643 when John Carmen and Robert Fordham bought a 60,000-acre piece known as Hemstead Plains. It was from here that the first livestock was driven to New York City and sold. Between 1860 and 1867 the founders of the town of East Hampton bought the last of what was called the "Lands of the Montauk" extending all the way to Montauk Point. In an interesting system rarely ever seen, the land was owned in common by locals that paid a fee to graze the common pasture. This unique arrangement continued mostly uninterrupted until 1879.

In 1879, East Hampton sold the land to Arthur Benson for $151,000. For the next ten years it was used in part as grazing land and for fishing and hunting. With the exception of 10 years before and during World War II, there have been cattle on the original "East End" pasture land since the first European settlers brought cattle to the island more than 350 years ago.

Today, the Deep Hollow Ranch sits near Indian Field south of Montauk highway. In the late 1970s, Rusty Leaver and his wife Diane bought the ranch to raise cattle and horses. Diane is the fifth generation on Deep Hollow Ranch and Rusty has been associated with the ranch since the sixties. Deep Hollow Ranch, originally started in 1937 by Phineas Dickinson, is now a part of the Montauk Park system and Diane and Rusty are the concessionaires for about 3,000 acres of county land offering trail rides and facilities for tourists.

The Leavers run a family orientated operation that focuses on good fun for everybody. They have entertainment for kids and adults as well as summer concerts, barbecues, and theatre. They also keep cattle and train cutting horses on the ranch, giving them a strong sense of family connection to the land, ranching, and the history of the island.

Today, if you could see the lighthouse and Deep Hollow Ranch from the Empire State Building, you would be looking at American history at its best, as well as some of the oldest cattle grazing land in the United States. Throw in a few pirates, some rum runners, and a cow or two and this long stretch of windswept coastline has a little something for everyone — even a die-hard Western cowboy.

Ghosts of the Wild West

"The Hold Up" C.M. Russell - 1889

To fans of the American West, a trip through Elko County, Nevada may feel as though you just took a wrong turn off the blacktop and drove onto a dusty old 1950's western movie set. If you pay close attention while you move through the sage flats and toward the distant Ruby Mountains, you can almost feel the presence of John Wayne as he mounts up and gives the signal to move out. Listen carefully to the sounds of the warm, swirling wind and you may hear the echo of the shots Randolph Scott fired at the rapidly disappearing stage coach robbers. Now, take it all in and close your eyes and you might even conjure up the ghost of Glen Ford stepping onto Main Street at high noon to face down the bad guy — like always, he's the one in the black hat.

A ride through Elko County, Nevada is like that — like a ride through your childhood memories. Cows vastly outnumber people and moving them and caring for them is the business and craft of a select band of rugged cowboys. Locally, they call themselves Buckaroos, and those that answer to that name seem to migrate to this part of the West, for it is Buckaroo central in the world of cowboy culture.

Elko County, Nevada is the fourth largest county in the continental United States with 17,000 square miles of land and only about 50,000 people scattered across its landscape. The city of Elko is the county seat and the heart and soul of Northeast Nevada. Today's cowboys and contemporary pioneers seek out this special place looking for that connection with the past where their mind can wander and they can find plenty of open space.

The earliest white man to see this country was the great explorer and mountain man Jedediah Smith, who first traveled through the area as early as 1826, and helped open the route to the Pacific Northwest. From the native Shoshone, Pyute, and Washoe Indians to the first white explorers, history paraded through Elko County non-stop. From 1842 on, the ox-carts filled with emigrants moved west with guides like Kit Carson looking for land to set down their roots. Within a few years, the gold seekers poured through the area heading for the California claims. John C. Fremont came to explore the Oregon Trail and the Great Basin country pursuing the United States dream of Manifest Destiny.

Travel in the early days was an ordeal and the early settlers and emigrants had to chose either a route along the Humbolt River or the Overland Pass over the Ruby Summit. Not every traveler reached his destination. The Donner party found themselves trapped when they reached the Sierra Nevada Mountains too late in the year and 39 out of 87 died during the winter. Few travelers thought of staying in the Basin, but those that did carved out a life that would soon turn into legend.

In 1868, those early settlers formed the town of Elko at the eastern end of the railroad track built by the Central Pacific Railroad (part of the Transcontinental Railroad). When the rail builders moved on, Elko became a service center for the growing mining and livestock industry. The economy grew quickly and the town soon became the hub for business and social life for nearly 300 miles in every direction.

By the 1870s, sheep had become another thriving industry in Elko County. The Basque people, a ranching culture from the Pyrenees Mountains area of Northern Spain and France, were the first to establish large bands of sheep. The Basque herders found they could take sheep in trade for their wages and soon hundreds of "tramp sheep outfits" as they were called at the time, were scattered across the public lands of Elko County. The Basque sheep industry has mostly faded into history now, but their sense of culture and pride is still celebrated in Elko every July at the National Basque Festival.

Elko County, Nevada sits almost entirely in what is called the "Great Basin" — a truly unique place containing mountains, rivers, and sage prairies. Called an "endorheic basin" by the geologists, all the water generated in the basin never reaches the sea. The largest river in the county, the Humbolt, is nearly 300 miles long and at its end, simply disappears into what is called the "Humboldt Sink." The Great Basin has become legendary in the West for both its uniqueness and its remoteness. For those who love western history, legends, and mystery, Elko County has more than enough to satisfy anyone.

The Humboldt Sink is not the only water mystery in the county. The story of a great ice-cold underground lake and river at the end of the Ruby Valley called Cave Creek even comes with its own in-house ghost. Legend has it that a small group of U.S. soldiers built a boat and hauled it to the small opening of the cave for further exploration. When they reached a wall preventing them from advancing, one of the soldiers dived down to see what was below. The survivors claim after five minutes, the rushing water spit out the body of the soldier — drowned and mangled. Some say that on entering the cave today, the ghost of that soldier still lives there — asking people to please go and leave him to himself.

What might be the last horse-drawn stage robbery in the U.S. happened in Jarbridge Canyon in Northern Elko County on December 5th 1916. Fred Searcy was driving the two-horse mail stage between Three Creek, Idaho and Jarbridge, Nevada when he was robbed of nearly $3,000 and murdered. The mastermind of the crime was one Ben E. Kuhl and two associates. What was unique about the case was a bloody palm print that was found on a letter at the scene. It marked the first time palm print evidence was ever used in a criminal case. Kuhl and cohorts were all sentenced to prison with the ringleader serving more than twenty-seven years.

In today's Elko, you can still walk down main street and rub shoulders with real cowboys, just look for the tall boots, spurs, and big hats. In the surrounding countryside, you can still see thousands of cattle and men on horseback, doing the same job they always have, and doing it pretty much the same way they have for a hundred and fifty years. If it's cows, Buckaroos, horses, and history you want, this is the place for you. Maybe you will catch a glimpse of the Duke driving cattle or hear the distant gunshots of the last stage robbery, or maybe you will just catch the spirit of the American West — how bad could that be?

The Fabulous 101

Miller Brothers Posters from about 1906

Once upon a time — 1881 to be exact, in a wild land filled with Indians, cowboys, and cattle, there was a man named George Washington Miller. Originally from Kentucky, he eventually landed in Northeastern Oklahoma in an area called the Cherokee Outlet and set up his own business. The story of the famous 101 brand is a little muddled in history. Some say it came from a local saloon and some say it stood for 101,000 acres of land. It is known that Miller first slapped the 101 iron on the left hip of his stock for the first time in 1881. From then on, the 101 became famous for just about everything that involved agriculture and western entertainment.

In the beginning, the old Cherokee Strip was largely unfenced and owned by the Indians. It could be leased for a few cents an acre and became a crossroads for the Texas cattle drives and a place where cattle could roam freely. The 101 raised horses, bison, poultry, hogs, and cattle — thousands and thousands of cattle. They also grew alfalfa, wheat, and corn, and experimenting with fruits and vegetables of different kinds. George Washington Miller died in 1903 and his wife Molly set her sons Joe, Zack, and George up to run the business. From this point on, the 101 became a household name in Middle America, not just for agriculture but for the best in Western entertainment.

Once promoted as the "greatest diversified farm on earth," the boys continued to search for ways to expand the use of their land and eventually got into the oil business. The profits from the oil set the stage for something radically new, even for the 101. In 1905, the brothers put together a genuine "Wild West Show." It rivaled anything of its day including "Buffalo Bill's Wild West Show," operating at about the same time. It was show business that got in the Miller boys blood and by 1907 the show hit the road and began to tour the country. From the beginning, it was a hit showcasing all the skills of its ranch cowboys.

The real hit of the show was the appearance of genuine American Plains Indians riding into the arena with war whoops and Geronimo, the famous Apache warrior who once killed a buffalo from the front seat of a moving car. One of the favorites was the famous Choctaw-Negro cowboy, Bill Pickett, who invented the sport of bulldogging. Life at the 101 was good, and as the success of the oil and show business grew, so did the diversity of the ranch. The Millers were now producing their own electricity for the operation and expanding their work in farming by experimenting with new strains of corn as well as apple, walnut, and pecan trees. The 101 became a self-contained city with stores, a tannery, and several different milling operations.

By 1916, the 101 took a break during the war years and re-formed in 1926 to tour again. The Millers took the "Wild West" not only coast to coast, but around the world to South America as well as Canada, Mexico, and even to Europe, performing for King George V and Queen Mary of England.

Most of the top western performers in the entertainment industry toured with the show in its day, including stars like Tom Mix, Mabel Normand, and Buck Jones. Will Rodgers was a staple at many of the touring events. At one point, Buffalo Bill Cody hooked up with the Miller show and toured with the "Buffalo Bill and 101 Ranch Wild West Combined." As the show continued to tour and recruit new stars, the industry of movie making began to come on strong. The Millers loaned out their cowboys and Wild West stars to the fledgling industry and many of them went on to successful careers in the business.

By the 1920s, the handwriting was on the wall for things like live western shows. As the movie industry grew in popularity, the box office for the 101 shows declined. In 1926, the 101 began to post losses they had never seen before. In 1929, the Great Depression took a dramatic toll on the operation causing even worse financial woes for the family. Joe Miller died in 1927 and George died two years later. The last remaining brother, Zack, could not stem the losses and by 1931, this incredible piece of western history passed away into the faded pages of American history.

What remained of the 101 was auctioned off. The last of the Miller Brothers, Zack, died in 1952. The fabulous 101 may be gone, but the stories of the 17-room concrete reinforced mansion called the "White House," built along the banks of Salt Creek, near Ponca City, still live in the memories of many. Generations of friends and family still talk about the Indians, the trick shooting, and roping shows. Pictures and stories of exotic animals like zebras, ostriches, and anteaters roaming the ranch still abound. A visitor to the ranch might see dancing mules and clowns, or be witness to stagecoach robberies and Indian attacks.

As the 101 began to fade, the refinery, tannery, powerplant, and the icehouse, as well as the repair shops and restaurant, went the way of the wild buffalo. The ranch even had its own novelty factory which produced beaded work, Indian rugs, and jewelry. You could buy your own hat, boots, and chaps and feel like part of the old west — at least in that part of Oklahoma. The 172 sections of land that comprised the old 101 are in different hands now, but if you visit there today, you can feel the history and visit the graves of Colonel Zack Miller, Bill Pickett, Trick Shot Artist Jack Webb, and many others are located in the area.

People in this part of Oklahoma just couldn't seem to escape the ghosts of the 101 and brought the "Wild West" back to Ponca City. This year celebrates the 56th year of the return of rodeo with the Ponca City "101 Wild West Rodeo." Fans of cowboy history and the Wild West can mingle with the spirit of the cowboys, Indians, and cattle that once called this place home. And maybe you will catch a glimpse of a buffalo or a longhorn, or maybe even the Ghost of Will Rodgers or Tom Mix riding drag in the swirling dust of the "101 Wild West Rodeo."

Ghosts of the Seeds-Kee-Dee

John Wesley Powell Ready to Shove Off into the Unknown - 1869

The small, bearded man stood upright in front of his chair, staring into the gaping entrance of the canyon. The Green River, a swirling, snarling maelstrom of thick mud, volcanic silt, and snowmelt disappeared between the towering vermilion walls. The one-armed man sat back in his chair, a custom built model strapped tightly to the deck of his twenty-one foot wooden boat called the Emma Dean. After one last look, he nodded his head.

It's June 9th 1869 and John Wesley Powell, with four wooden boats and nine fellow explorers, shove off into the unknown cataract and on to the pages of American history. Before they enter the unknown gorge, they name it *The Canyon of Ladore,* a name suggested by an old poem. By the time they completed their journey, they had floated the Green River from Wyoming to the confluence of the Grand River (now the Colorado River) and on to the end of the Grand Canyon.

The Canyon of Ladore rises dramatically above the river in a very special part of the world called Brown's Hole. The canyon is in Colorado, but Brown's Hole itself is a wide flat valley thirty-five miles long by six miles wide, shared by Utah and Colorado and brushes up against Wyoming on the North. The Green River meanders the length of the valley and leaves by way of the red walls of the Canyon of Ladore.

By the time Powell made his epic exploration of the Green, this special place was already steeped in history and legend. The Indians lived in the isolated valley for hundreds of years until the first Spaniards appeared in the late 1500s, building a fort near the mouth of the canyon. Legend says the Indians massacred the occupants and burned the fort about 1650. The Kohogue Shoshone nation called the valley *"O-Wi-U-Kuts"* meaning Big Canyon and called the River *"Seeds-Kee-Dee"* or Prairie Hen.

It was 175 years before white men were seen again in Brown's Hole. These men wore fringed buckskins and searched for Beaver. The first records of the American Fur Trappers show William Henry Ashley brought in a party of trappers about 1825. Ashly and his men floated the river into the Hole and camped "on a spot of ground where several thousand Indians had wintered. Many of their lodges remained as perfect as when occupied." Ashley and his small band of trappers drifted past the shore near where the early Spaniards built their first fort. They floated into the unknown canyon in the spring of 1825 in homemade buffalo skin and reed boats called 'bullboats.' Within a mile, the trappers encountered the violent rapids that would claim Powells boat the 'No Name' 39 years later, a place he would name *Disaster Falls*.

By the 1830s, the trappers had built a rude log and stone fort they named Fort Davy Crockett. From this site, the trappers established a trading business with the local Indians. By the early 1840s, the fashion of Beaver hats had all but disappeared and the Beaver-trapping business fell on hard times. By the mid 19th century, Brown's Hole had seen all the great mountain men like Kit Carson, Jim Bridger, Joe Meek, and Jedediah Smith. The time of the mountain men was passing, and settlers with new ideas began to move into the sheltered valley.

From the West end of the valley in Utah, at the beautiful Flaming Gorge, to the Southeast end at Ladore Canyon in Colorado, the Green River made its way wandering through lush grassy fields and Cottonwood groves. People realized that this protected valley was a natural for grass fed livestock and the cattle industry slowly began to fill up the voids left by the Indians, trappers, and explorers that had called this place home for hundreds of years.

In 1847, a fiery young Mexican cowboy named Juan Jose Hererra, a man with a dubious past, convinced his brother Pablo and a group of like-minded Mexican vaqueros, to gather a herd and take them to Brown's Hole. The cowboys left California, and by the time they reached the valley, they had a herd. How they accumulated the cows was the subject of much conjecture, but Brown's Hole now had her first resident herd of cattle. After moving the herd over the remains of Fort Davy Crockett and past the Canyon of Ladore, they stopped at the base of 'O-Wi-U-Kuts' Mountain and built their 'rancho.'

In 1848, Brown's Hole was in the land ceded to the United States from Mexico and as the West opened up, various immigrant trains and cattle drives stopped to winter in the valley. After the Civil War, herders trailed Longhorns from Texas through Brown's Hole and on to California — trips that often took up to two years. The valley, though still remote, became known as a safe place to set up winter camp.

With the expansion of the West came settlers. Family names like Crouse, Hoy, Warner, and Basset are still carried on the canyons and streams today. With the prosperity of the cattle ranching settlers came another unique chapter in the valley's history — outlaws. Around 1886, a young, fair-haired cowboy came to work for rancher Charlie Crouse, he said his name was George Cassidy. Cassidy proved an able hand with horses and easily made friends with Crouse and many others in the valley.

These friendships would save his neck many times in the next few years as Cassidy, now known as 'Butch,' formed his train and bank robbing gang. Brown's Hole, by now called Browns Park by most people, proved to be the safe haven needed by the outlaws. Cassidy, along with gang members like the Sundance Kid and Kid Curry, used the park as a base, forming a loose alliance of outlaws known as the 'Wild Bunch.' The outlaws met regularly to discuss business, play cards, and hold horse races. Friends made in Browns Park kept an eye out for them and tipped them off when the law was in the area.

Cattle also meant rustling, and the stories of rustlers like Anne Basset, 'Queen of the Cattle Rustlers' are legend. Tom Horn is said to have plied his trade here and the park is home to outlaw hideouts, lynching, gunfights, and even a Diamond Mine — salted by unscrupulous miners. The South end of the Canyon of Ladore, where Ashley and Powell exited the canyon, is today the site of Dinosaur National Monument.

They say if you camp anywhere in Brown's Hole you are camping directly on the history of the West, so tread lightly. The ghosts of Indians, mountain men, explorers, and outlaws jealously guard this special place where the 'Seeds-Kee-Dee' river cuts a deep path through time and history.

The 200-Year Range War

"To the Victor Belong the Spoils" C.M. Russell - 1901

To listen to the cowboys of the lower 48, raising and caring for their cattle can be a miserable and often dangerous proposition. Heat, cold, drought, flood, and rattlesnakes are some of the common complaints heard throughout history. Stories of knot headed horses that blow up at exactly the wrong moment fill page after page of the legends and stories about cowboys. Musicians, writers, and poets have made a living for more than a century telling the world of their trials and tribulations. Of course, the cowboys are right, it has always been a tough and dangerous way to make a living. However, there may be a few cowboys that have it even tougher than the southern hands — and a whole lot scarier.

By comparison, the beef cattle industry in Alaska has never been large, and what there is of it, is scattered all around our largest state, with a lot of it taking place on the southern coastal islands. Kodiak Island was first settled in 1763 as part of Russia, shortly after that it saw its first bovines. As luck would have it, the island's natives also had a taste for fresh beef. From the start, the ranchers had to fight to keep these local poachers away from their stock.

Poachers anywhere can be a problem, but the poachers on Kodiak Island were 10' tall, 1500-pound Kodiak bears — possibly one of the meanest critters on the planet. After Russia turned over Alaska to the U.S. in the sale of 1867, entrepreneurs from everywhere came to Kodiak Island and other areas to stake their claim on a future in Alaska. The more people that arrived looking for their fortune, the greater the need for fresh meat.

The giant bears greatly appreciated the efforts of the early settlers to raise livestock and quickly became the cattleman's best customers. As the battle between the bears and the ranchers heated up, the early settlers tried everything they could to keep the giant carnivorous bears (actually omnivores) at bay. They shot every bear they could find, and it appeared to make little difference in their numbers. They brought in large, aggressive dogs in an attempt to scare the bears off, but most ended up as a snack for the ever-hungry cattle poachers.

At one time, there were more than a dozen ranches doing business on the rolling, well-watered hills south of the town of Kodiak. The ranches averaged about 22,000 acres and were the result of the 1887 Hatch Act that set in motion the opening of federal lands for the use of agriculture.

By the 1950s, one of the legendary Kodiak ranchers of the time, a man named Joe Zentner, decided to step up his game and buy his first airplane, a Piper Cub purchased from a dealer in Kansas. Zentner paid $3,800 for the plane and $1,200 to have it delivered to the island. When he finally got it home, he still had to learn to fly it, and another Kodiak rancher and pilot, Dave Henley, taught him the basics of flying. After a few flights with Henley, Zentner, impatient to find some bears, took off by himself. He flew the rest of his life without the benefit of any more lessons or a pilot's license.

Alaska was still a territory, and hunting bears from an airplane was a gray area in the law at that time, but it didn't deter Zentner and his friends from the mission at hand. It also didn't slow down the bears very much. Always looking to be more efficient, Zentner mounted an M-1 semi-automatic rifle on top of his plane. The rifle fired a mere four inches above his propeller and as soon as it was secured, he took off in search of his old adversary.

Until 1963, the flying bear hunters (by now there were at least three) continued their bear strafing runs pretty much unabated. By now, Alaska was a state and when the residents of Kodiak saw the planes equipped with guns, the news got out and Alaska's Governor Egan ordered an immediate halt to the flying killing machines. Although Zentner still made an occasional sortie over the ranch, it all ended when the August, 1964 Outdoor Life Magazine hit the stands with a story called "The Kodiak Bear Wars," complete with cover art depicting a plane blazing away at a giant Kodiak Bear standing on its hind legs raging at the plane.

The next great idea was the proposal for a 9' steel fence to be installed around the ranching areas. The fence would run for miles and miles and the cost would be astronomical. Since the demise of the flying "Bear Force," the bears were on the increase, and, like always, still held the upper hand. After kicking around the fence idea with all the ranchers, and state and federal officials the fence idea died a natural death. The cattle industry finally realized that when it came to the cattle, or the bears, the bears would always win out.

For nearly two hundred years the battle was fought by every means at hand, by the showing the humans just who was toughest kid on the block. The records show that in 1927, there were about 1,000 cattle, both beef and dairy combined, on the Kodiak and Aleutian Islands. Estimates from the mid-1960s only added up to about 1,300 head total.

There are still a few ranches left in the traditional bear areas, run by people nearly as tough as the bears.

Today, the livestock industry in Alaska is a modest, but steady business. It now includes reindeer, yak, sheep, and hogs as well as cattle. On some of the old ranches, bison have replaced cattle and are generally considered to be better in the harsh environment and less vulnerable to the bears.

As for their southern counterparts, they like to point out that at least they are never too hot and they don't have rattlesnakes up there. They're right of course — they just have 1,500-pound eating machines always looking to poach a free meal . . .

Hooves, Horns, and Grass

Part-1

"When the Land Belonged to God" C.M. Russell – 1914

"I ascended to the top of the cutt bluff this morning, for whence I had the most delightfull view of the country, the whole of which except the valley formed by the Missouri is void of timber or underbrush, exposing to the first glance of the spectacular immense herds of Buffaloe, Elk, Deer, & Antelopes feeding in one common and boundless pasture." Meriwether Lewis, April 22, 1805.

What Lewis saw from that bluff in western North Dakota on a cool April morning, was the edge of one of the largest wild grasslands in the world. The early French trappers and explorers of North America called it the "prairie," meaning meadow, and the name stuck. Worldwide, it is comparable to the South American Pampas, the Serengeti of Africa, or the Steppe of Asia.

Thousands of square miles of nutritious grass, forbs, and shrubs drew these animals to the prairie. All of these have three important characteristics in common that made it a match made in heaven — they're all herbivores, all ruminants, and all ungulates.

This vast area of prairie stretched from Saskatchewan, Canada, to central Texas. Considered to be three distinct zones — from the Rocky Mountains going east there is the shortgrass prairie, the mixedgrass prairie, and the tallgrass prairie.

The formation of the North American prairies began to take place with the upwelling of the Rocky Mountains. In prehistoric times, as the mountains started to push their way up, they created a rain shadow that slowly cut off the moisture to the east and caused most of the trees to die off and the terrain to dry out. The land closest to the mountains received less moisture and the grass came in shorter. The prairie land farthest to the east had the benefit of more rainfall and became the tallgrass prairie. Average rainfall on the prairie is considered to be 10" to 30" a year.

After the receding of the last ice age, approximately 10,000 to 15,000 years ago, the earliest Native American ancestors started their migration to North America across the Bering Land Bridge. Among other early arrivals to the continent was the bison, and as the herds began to increase, they started to follow the fresh green prairies south. In time, the mating of these giant grass eaters and the large areas of wild prairie grass slowly created their own self-contained ecosystem.

This magnificent, shaggy, eating machine we call buffalo, grew into enormous herds, grazing their way through thousands of acres a day and moving steadily as they ate. As they chewed the grass down, they passed the seeds through them and their hoof imprint loosened the soil and pressed them into the ground, helping to replant as they went. In time, they would become one of the most important animals in the American West.

Even before the arrival of horses, the Native Americans living on the prairies hunted them on foot. Archaeologists have excavated the remains of buffalo with stone spear points found between the ribs, the type that were used thousands of years before white men walked the continent. Carefully chosen canyons and cliffs were often used as places to stampede them over the edge as a method of collecting as many as possible at one time.

When the Native Americans mastered the use of horses, the buffalo became even more important — they became part of their culture and religion. The horse allowed them to kill only as many as they needed and only the animals that were considered the best. They consumed every part of the animal, including the intestines, the brains, and the marrow. The hides became lodges, clothing, and items for trade.

By the time the white settlers started to gain a foothold in the Eastern United States, this denizen of the wild prairies numbered in the millions. Single herds often numbered in the hundreds of thousands, if not a million or more. Some estimates put the North American Plains Buffalo population at thirty to seventy million as recently as two hundred years ago.

The meeting of the buffalo and the grass proved to be a perfect moment in the history of North America's prairies. This 2,000-pound, grass loving, cud chewing ruminant with the cloven hoof of an ungulate not only flourished, but helped the grass to grow and provide a place for many other species to reach strong, healthy populations.

After the turn of the nineteenth century, the early explorers and trappers discovered that a new market for tanned hides was developing in Europe and in America. The tough hide was also perfect for many other products, and every industry that used leather in any form, from the shoe and saddle industry to industrial pulley belts, clamored for more hides.

Between 1820 and the start of the Civil War, professional white hunters and Indian hunters killed an estimated three million animals a year with little serious effect to the herd size. Most of these were killed just for their hides, the carcasses left to feed the various predators.

After President Lincoln signed the Homestead Act on May 20, 1862, the stage was set for the greatest rush for free land in history. A few months later, he signed the Pacific Railway Act, an action that would forever change the history of the greatest icon the American West had ever produced.

Several wagon trails through the grasslands had been forged by pioneers in wagons from St. Louis to places like Oregon, Santa Fe, and California, and were used for decades. For some unknown reason, in their haste to get to the Pacific Coast, these early settlers appeared to pass through the prairies with blinders on. They saw little value to the vast expanses of grass, considered by many to be a desert. Instead, they hurried west to stake their claims and start their farms and ranches in Oregon and California.

May 10, 1869, marked two important moments in the history of the West. On that day the last spike was driven at Promontory Summit, Utah Territory, finishing the rail line — and at the very same moment marked the start of the death march of millions and millions of the grasslands most perfect resident — the American Plains Buffalo.

With the railroads came the hide hunters in unprecedented numbers, they could now kill and ship tens of thousands of hides a week. Some hunters were to become legends, killing thousands in just days or weeks by themselves. U.S. General Phil Sheridan promoted the extermination as the easy way to remove the Indians from their ancestral lands. Even though there were laws regulating hide hunting, the government turned their head at the shameless slaughter of these magnificent animals. By the 1880s, the buffalo and the prairie ecosystem it helped create was doomed to be a footnote of history.

Hooves, Horns and Grass

part-2

Taking the Tongue – photo by L.A. Huffman about 1870

By 1890, the story of the American plains bison (buffalo) was all but over. The tale had started with tens of millions of 2,000 pound animals and finished with under a thousand captive animals in less than a hundred years. Man, like always, continued to prove there were few things he could not conquer. The twenty-five years after the Civil War had proven to be devastating to the American prairie. Millions of buffalo were wiped off the face of the earth, mostly for their hides. Their rotting carcasses were left scattered across the grass from Texas to Canada.

At the same time, the plains tribes lost not only their major source of sustenance, but their very identity. By the end of the century, most plains Indians had been moved into reservations and the Indian and the buffalo, whose lives were so intertwined for thousands of years, were, for all practical purposes, just a footnote in the history of the American prairie.

The railroads had laid their steel across the West, and thanks to the Pacific Railway Act, found themselves the sole owners of millions of acres of land in the now empty prairie. They set up towns and stations along the way, selling the remaining land to anyone who had the money. Wealthy people from all over realized that this was a huge opportunity to get into the live cattle business. Many of the early ranches were funded by investors from England. The common joke among the Americans at the time was this was where the English fathers sent their errant sons to keep them out of trouble.

The loss of the buffalo caused a domino effect throughout the prairie ecosystem, the rest of the ruminants that shared the prairie like the elk, deer, and antelope became harder to find. Millions of hooves no longer softened up the ground and deposited new seeds. Predators like the wolf and coyote began to disappear, even the smaller prairie scavengers found it hard to make a living in the newly barren country.

The prairie ecosystem of the last ten thousand years had been severely damaged, and would continue to be abused for many more years to come. For many years, the cattlemen had been bringing in Longhorn cattle from Texas. These tall, lanky, wild cattle had been raised in some of the most inhospitable country Mexico and Texas had to offer, and had learned to survive almost anything mother nature could throw at them.

The Longhorns also came with a built-in problem, they carried Texas Fever, a disease spread by cattle ticks. Over time, they had become immune to the disease, but wherever they went, the local cattle would pick up the diseased ticks and suffer a miserable death. The early ranchers, men like the cattle barons of Wyoming, found that many states and places along the new rail line had started to ban moving the Texas Longhorns across their states.

The members of the Wyoming Stockgrowers Association, the same group that would be later responsible for the famous Johnson County cattle war, realized they needed to adjust their thinking on the cattle they raised. By the middle eighties, led in part by the influence of the English investors, European shorthorn breeds like Angus and Hereford (then called blooded cattle), were imported and the Longhorn rapidly disappeared from the open range.

Replacing their existing cattle with the European breeds was considered by most to be the right thing to do, but the shorthorns were much slower and more docile, and proved to bring with them their own set of problems. They were not by nature grass eaters like the buffalo or even the Longhorn.

They had evolved in a part of the world that had more than enough rainfall, and had evolved into very picky eaters. They preferred plants of the leafy variety, searching out broadleaf plants and forbs over the regular grasses of the prairie.

The shorthorns would eat grass when necessary, but would always migrate to the streams and areas where they could find an abundance of leafy, green, broadleaf plants. The buffalo had loved the wide open prairies full of wild grass, and helped it to flourish. When the winters gave them a lot of snow and ice, they would paw the ground and swing their massive heads back and forth in the snow to reach the grass underneath.

The European cattle simply hadn't evolved to survive the winters in the American West. They would graze the ground to the dirt before they went to the grass. When the blizzards hit the prairie, they would be knotted up in the gullies and the streambeds, found in the spring — dead by the thousands.

By the 1890s, the days of the open range style of cattle ranching was over, the range wars had fallen to the flood of small farmers and ranchers and the hated barbed wire. It had really died years before, but many of the early ranchers refused to let it go — the idea of getting rich on the open range was just too good to give up.

In time, the cattle barons of the day began to catch on to what a future without free public land for grazing might look like. They soon started to buy all the land they could and stocked it with the shorthorn breeds. Large private ranches were born, and on the fringes, the farmers with their quarter-section and half-section claims began in earnest to break the sod of the American grassland.

Left mostly unregulated, the prairie suffered the indignity of having more than forty-percent of the grassland seriously overgrazed, and even more of it wasted to bad farming practices. By the early 1930s, the government finally woke up to the degradation of the Western grasslands and passed the Taylor Grazing Act of 1934. It set up the system of grazing allotments we still have today.

Looking back to that day in 1805, when Meriwether Lewis recalled the *"immense herds of Buffaloe, Elk, Deer & Antelopes feeding in one common and boundless pasture,"* we have to wonder if he could have ever imagined what this magnificent ocean of grass might come to in the short span of a hundred years — probably not. Could he possibly have imagined tens of millions of buffalo turning into a few hundred in the same short span? Again, probably not.

With the foresight of several people in the late nineteenth century, a campaign to preserve the American Buffalo began to slowly bring back America's favorite icon. Thanks to those forward-thinking people, the greatest icon of the American West was saved. The estimates of the populations today are about 500,000 animals. As for the historic prairie grass, what's left of it is preserved in several locations around the West and there are now more than enough buffalo herds around the country for everyone to see the greatest American icon of all.

The Wild Owyhee

"Infantry Attacking the Snake River Indians" — Arthur Wagoner - 1880

The term 'Wild West' has been used in conversation, in the movies, and in writing of all kinds for at least a hundred years. At times, everything west of the Mississippi River has been called the Wild West, from the flatlands of the Midwest to the shores of the Pacific. After a hundred-and-fifty-years of western expansion, what we think of as wild has gradually been tamed into cities, towns, ranches, farms, and people — lots and lots of people.

Often, people who live in the East and in the big cities are heard bemoaning the loss of the wide-open spaces of the old days. They love talking about the fact that the Wild

West and all the real cowboys are gone, and how they would have loved to live a hundred-years ago so they could have been part of that life.

Their facts are partly right. The West may not be as wild as it was back then, and maybe there aren't as many wide-open spaces as there once were, but they're out there if you want to look for them. If you really want to see some of those places you've heard about or read about while growing up, you may be required to get a little way off of the pavement.

On March 4, 1863, President Lincoln signed the Idaho Territory Act into law. It broke up the politically disputed Washington and Oregon Territories and initially made the Idaho territory large enough to include most of present-day Wyoming, Montana, and Idaho. The first county in the new territory was called *Owyhee County* and was situated in the extreme southeast corner of the new boundaries with the Oregon and Nevada territories.

The new county had been called the Owyhee since the early trappers of the Montreal-based Northwest Company explored it. The name came from three Hawaiian natives that joined the trapping expedition of 1819. The natives of the islands had been dubbed *Owyhee's* by Captain Cook after he first set foot on the Sandwich Islands in 1778, and the name stuck. When the three natives mysteriously disappeared exploring a fork of the Snake River, the rest of the expedition named the region the Owyhee for the missing trappers, and the name stuck again.

The northern most edge of the Great Basin consists of rolling sage prairies, deep canyons and mountains, and is cut through by the Owyhee River. Bordered on the north and east by the Snake River, the famous Silver City mining area lies in the Owyhee Mountains in the north of the county. The famous Dry Route of the Oregon Trail cut through Owyhee County, becoming the first real road in the region and was used as a trail for the immigrants for more than thirty-years.

After gold was discovered, the boom brought in thousands of miners searching for their piece of the treasure. Like most gold rush towns of the period, some found gold, but most found failure. However, successful or not, they all came hungry, and beef in this isolated region was almost non-existent.

The first large herd of cattle finally reached the new territory of Idaho and entered the county through the Bruneau Valley in the fall of 1869. For nearly a year, a handful of men from Owyhee County struggled through the unpredictable weather and imposing western geography to drive a herd of 1,400 cattle up from the Brazos area of Texas. Most were Durham cattle, with a few Longhorns mixed in. These would become the seed stock of the new Idaho cattle industry, and the hungry miners and citizens of this remote outpost couldn't wait for their first steak.

In a few years, Owyhee County had an estimated 100,000 cattle and many ran free, much like their early Longhorn counterparts in Texas. As the mining boom began to wind down to its inevitable bust, ranching and farming began to become the next big industry. In a few years, the region became prosperous as the cattle and sheep industry boomed with tens of thousands of animals fattened up on nearly unlimited grass and driven to the nearest railhead in the fall.

Even with the success of the livestock industry, the life in Owyhee County was difficult, its extreme isolation making day to day life a lonely struggle for most of its residents. Ranches were often so large, and so far from civilization that people seldom left because the travel was too difficult.

Agriculture in general remained a tough life until 1864, when congress passed the Carey Act, also known as the Federal Desert Land Act. It allowed private entrepreneurs to build irrigation systems on federal land and sell the water to ranches and farms. The soil of the Owyhee region is great for agriculture and will grow nearly anything, but needs more water than is produced naturally.

With the water came the next big boom, farming the dry lands of Owyhee County. Today, most of the farm land of Southern Idaho is heavily irrigated and produces millions of bushels of potatoes, corn, beets, and wheat, as well as crops like alfalfa.

Today, Owyhee County is still one of the largest in the U.S. and one of the most remote. The corner of Idaho, Oregon, and Nevada, and the Great Basin region remain a place for those who like wide open spaces, and peace and quiet. And, if you are serious about finding the real Wild West, take a drive down Interstate 80, then turn north, somewhere around the famous old cowtown of Elko, Nevada, and drive for a couple of hours. If the empty Nevada landscape doesn't already look a little like the Wild West to you, cross through the Duck Valley Indian Reservation and into Owyhee County, Idaho. Today, the region has a population of 11,500 people scattered across 7,676 square miles for an average of 1.4 people per square mile.

This is the land of wide-open spaces, the land of the cowboy and the cattle, and a place to find some of the real Wild West. Genuine working cowboys saddle up for work every day, rope cattle and build a fire to heat up their irons for the next branding. They still gather the cattle in the fall and drive them to market. They're all out there, somewhere in the wide-open spaces of the Owyhee.

Kittitas County

"Trouble on the Horizon" C.M. Russell - 1893

Tucked between the Columbia River and the Cascade Mountains, a hundred or so miles from Seattle, is the Kittitas Valley. The earliest settlers of this valley were the Psch-wan-wap-pams, (stony ground people) considered part of the Yakama tribe. Up in the northwestern corner is the Wenatchee National Forest and the Manastash, and the Umtanum ridges form a natural border between Kittitas and Yakima counties. The Yakima, Teanaway, and Cle Elum Rivers also wander their way through the county.

In case you haven't had enough fun with local native words yet, nobody appears to be positive exactly what Kittitas means. It's pronounced *Kitt-i-tass,* and most seem to think it refers to the ground, or the white clay. The Columbia River became the east boundary of the county and the historical home for dozens of Wanapum Indian villages. The first European to visit the area is thought to be a French-Canadian fur trapper, Alexander Ross, about 1814. He described the scene he saw in *Fur Traders of the Far West:*

"This mammoth camp could not have contained less than 3,000 men, exclusive of women and children, and treble that number of horses. It was a grand and imposing sight in the wilderness, covering more than six miles in every direction. Councils, root gathering, hunting, horseracing, foot-racing, gambling, singing, dancing, drumming, yelling, and a thousand other things which I cannot mention, were going on around us."

Ross calls this place the Eyakema Valley. Historians now identify it as the Kittitas Valley. In 1859, a cattleman by the name of Ben Snipes drove the first herd of cattle through the valley, past the future home of Ellensburg, all the way to the newly opened goldfields in Canada. Snipes and his crew also drove cattle through the area to Tacoma, Seattle, and British Columbia. The Kittitas Valley was a well known location, favored for its good water and plentiful grass and became a popular place to rest the herds before they pushed on over the Cascade Mountains.

After a wagon road was built over Snoqualmie Pass in 1867, the valley began to grow into a ranching center. Enterprising businessmen took up the open land and the cattle industry soon became the economic engine that drove the area.

Soon after the first immigrants moved to the valley, missionaries of the Catholic Church, in their never-ending crusade to civilize the tribes, began to establish a presence in the area, building the Immaculate Conception Mission in the valley in 1847. Like many times in the history of the American West, this would begin the slow decline of the great Indian nations of the area.

When the Washington Territory was established by Congress in 1853, the new governor, Isaac Stephens, took his assignment seriously, and began negotiating with the tribal chiefs for ownership of nearly 17,000 square miles of historical Indian land, or nearly one-fourth of the new territory. The tribes signed the treaty on June 9, 1855. They were left about 1,900 square miles for their reservation. Shortly after the signing, local papers declared that the freshly ceded land was now open for settlers and entrepreneurs of all kinds. With the flood gates now open, the gold miners poured into the county after word of a strike on Swauk Creek, followed by the discovery of large coal deposits.

As the valley became overrun with gold seekers, the conflicts with the local Indians became commonplace, and it eventually led to the Yakima Indian Wars in 1855-1858. Not all the tribes were happy with the agreement, but the Army forced the angry tribes back onto the reservation at Fort Simcoe to keep the new settlers safe. One of Kittitas Valley's earliest settlers, cattleman, and prominent state senator, A.J. Splawn, wrote about selling horses to the miners.

"By 1864 this part of the country was gold-mad . . . Spreading out like a fan, the gold hunters invaded every hole and corner of the mountains. Thousands of cattle, driven in from the lower Yakima summer range, grazed the beautiful valley whose fine bunch grass grew even up to the water's edge. There were no flies of any kind to disturb the stock and there was cool, clear water in numerous small streams that wound through the grassy plain. The cattle became so fat that they had to hunt shade in the early morning. It was a veritable cattle heaven . . ."

Splawn and his partner Ben Burch opened the *Robbers Roost Trading Post*. After a few years, they sold it to newcomers John and Mary Ellen Shoudy. The Shoudy's were instrumental in founding the town, naming it Ellensburgh. In 1894 the U.S. Post office dropped the "h" in the name and it officially became Ellensburg.

The Northern Pacific Railroad reached the valley in 1887, and was the final connection needed to move Kittitas County and their beautiful valley forward into the new century. Farming began to gain popularity and dairy farms began to spring up. Improved irrigation systems caused a dramatic increase in wheat and hay production. Vegetables like potatoes and sweet-corn became cash-crops as well as fruit trees.

On November 11, 1889, Washington formally became a state. By this time the population was mostly immigrant whites and the very nature of the valley had changed forever. *The Tenaway Bugle*, a small-town valley paper published an article from one of the early settlers:

"Less than a year ago, Tenaway City was a howling wilderness while today we have a healthy village with a bright prospect for the future."

From the 1850s to 1889, Kittitas County, Washington went from the real Wild West to the 42nd state in the union. The place is settled now, full of hay fields, cattle herds and fruit trees. Gone are the thousands of Indians and their herds of ponies. In their place is modern civilization with some of the most interesting place names in the West. A.J. Splawn fondly remembered back to the early days in his writing:

" . . .Commercial crazes and get-rich-quick schemes had not yet reached this wild beautiful land. The people were honest and happy. They sold their cattle once a year, and consequently paid their bills only once a year, but the trader knew he would get his money."

The Last Squeal

Chicago Union Stockyards – about 1900

Imagine for a moment 475 acres of hogs, cattle, and sheep all in one location, in the middle of your town. Add in more than 50 miles of roads and 130 miles of train tracks inside the enclosure to service all this stock. Then, throw in more than 25,000 employees at its peak and you have a quick idea of what the Chicago Union Stockyards were like for more than a century. Stretch your imagination just a bit further and you can almost see — or more accurately smell — the assault on your olfactory receptors! Chicago, by the turn of the century, may well have been the smelliest place on earth and quite proud of it.

The story of the Chicago meat packing business is an extraordinary slice of the American story, taking us from the Civil War all the way to 1971, when all but the highly ornamental front gate was finally torn down. On Christmas day 1865, the "Union Stockyard and Transit Company" opened her gate to the cattle, hog, and sheep industry and by 1900, the Chicago Stockyards were producing 82% of all the meat consumed in the United States. The cast of characters in the stockyard story includes thousands of immigrants from dozens of countries imported to do the dirty, dangerous, and backbreaking work. Strikers, strikebreakers, famous journalists, and no less than President Teddy Roosevelt himself, all played a part in the formation of the live animal processing business as we know it today.

By the turn of the century, the Chicago operation was slaughtering tens of thousands of cattle, hogs, and sheep every day. In the hog operation alone they had over 2,000 pens holding as many as 75,000 hogs on site at any one time. The self-bestowed title "Hog Butcher of the World" proved to be accurate and they continued to slaughter hogs at a mind-numbing pace well into the twentieth century. On one record day, they claimed the all-time hog kill record at 190,000 in twenty-four hours.

The stock was killed and butchered with an almost ruthless efficiency. Every part had a use: from the meat, to fertilizer to exotic items like violin strings, not one single thing was wasted. Phillip Armour, owner of the largest packing operation of the day, famously remarked that when they finished with an animal, "nothing was left but the squeal." Legend has it that after touring the yards and watching the efficiency of the operation, Henry Ford got the idea for his first assembly line.

After the turn of the century, the operation was known for more than just meat production, it also became famous, if not infamous, for the abysmal treatment of its workers and handling of the meat. By the time the yards started operation, all of the large meatpackers of the day had begun to set up shop. Names we know today like Swift and Armour as well as others like Morris and Hammond took advantage of the location and the water resources.

The operation took more than half a million gallons of water daily from the nearby Chicago River and returned it to the river mixed with tons of waste every day. The name "Bubbly Creek" was given to the fork of the river receiving the unholy mixture and it became just one more black mark against the operation. By 1872, the industry had started working with ice cooling, and by 1882, had developed the first refrigerated railroad cars, which allowed full production summer and winter.

If working in the yards was difficult and dangerous, it was still not as bad as the killing operations. Workers on the kill floor suffered through freezing winters and sweltering summers often standing in ankle deep blood, excrement, and water for ten or twelve hours a day. The wages and benefits were poor even for the time, and employers knew that there was a nearly endless supply of immigrants waiting for any work they could find. Chicago's population rose, and became as diverse as places like New York City. The German and Irish were among the earliest on the scene and opened up the floodgates for almost every European, African, and Mexican worker to follow.

In 1906, a journalist named Upton Sinclair published a fact-based fiction novel that turned into a staggering expose' of the Chicago meatpacking industry called *The Jungle*. The world for the first time got a look inside of the animal slaughter business — too good of a look as it turned out. His account of workers falling into meat processors and being ground up along with the animal trimmings stunned the country, if not the world. Foreign sales of American meat fell by an estimated fifty-percent and the producers knew it was time to start cleaning up their act. The major producers themselves in an attempt to calm the outrage lobbied Congress for new controls and inspections in their own plants.

Through prodding by Sinclair and others, President Theodore Roosevelt finally called for an investigation into Sinclair's claims. What he found was appalling by any standards, the contamination of the product, and the treatment of the workers was finally coming to light. Sinclair and Roosevelt were instrumental in passing the *Meat Inspection Act* and the *Pure Food and Drug Act* of 1906.

Although Sinclair's book helped the issue of healthy meat products, it did little to address the plight of the worker. The owners of the slaughterhouses and the stockyards fought the labor movement through many rounds of attempted strikes. They brought in thousands of strikebreakers and successfully held back the organization of the workers for many years. It wasn't until the *Labor Relations Act* of 1935 that workers were guaranteed the right to organize and Congress outlawed many of the historical practices the owners used to keep out organized labor. From that point on, the industry began to slowly clean itself up.

That the world became aware of so many things through Sinclair's metaphorical novel is obvious. That the world we live in has improved so dramatically from it, is not always obvious to many. Today's clean meat, water, improved working conditions, and diverse population can be, at least in part, traced to the Stockyards and the meatpacking industry in Chicago.

Today, the yards are completely gone from the landscape, the only memory of it is an ancient limestone gate that served as the entrance for a hundred years. On July 31, 1971, the Chicago Stockyards finally heard its last squeal.

Highway of Hooves

"Cattle Drive" C.M. Russell - 1898

In January of 1885, southwest New Mexico and eastern Arizona was a remote, rough and tumble place, full of wild cowboys with fat cattle and no one to buy them. The difficulty of life in this part of the country was dramatically compounded by the distance of the cattle from potential buyers.

The history of the Western cattle drive has been well covered over the last century, and trails like the Chisholm and the Goodnight Loving stand out as two of the most famous. The truth is, most of these early trails rode out of history nearly as fast as they rode in, seldom lasting more than a few years. Railroads became the death blow for most old style trail drives. Moving cattle to markets hundreds of miles away now meant moving them to the nearest rail spur. As the old trails faded into legend, there were a few places in the West that still remained beyond the reach of the rails.

By 1880, the Atchison, Topeka, and Santa Fe had already established a line from Colorado through the New Mexico Territory to El Paso, Texas. At the urging of the local livestock industry, they agreed to build a spur from Socorro west to Magdalena. The world of the cattle and sheep industry in Southwest New Mexico was suddenly presented with a permanent place to bring their stock and the *Magdalena Livestock Driveway* was born.

Even today, this part of the world remains a remote, rough, and tumble place. It's still full of cowboys (maybe even a few wild ones) with fat cattle to be shipped. Although today's operations don't have to depend on long drives to get their product to market. In 1885, when the rail spur finally hit Magdalena, New Mexico and Eastern Arizona saw light at the end of the cattle drive tunnel.

Springerville, Arizona, founded in 1879, was the initial starting point of the trail. From the beginning, the trail was open range and used continually until the new *Stock-Raising Homestead Act* of 1916. This act allowed much larger claims for homesteaders. They could claim one section of land for ranching purposes with a minimum of land improvements. Many of the best claims were along the cattle driveway. The new homesteaders would own the surface rights and the government retained all of the subsurface rights. Livestock breeders using the driveway for generations claimed that this would damage or even block the historic trail and petitioned the Secretary of the Interior to exclude the trail from the act. The government complied with their request and removed the trail from the Homestead Act. The Magdalena driveway was now official and protected by law.

The trail was mapped out to average at least five miles wide and as much as ten in some areas. Even as wide as it was, range abuse was common. Forage was always at a premium, and the local cattle ranchers and sheep operations often left little for those coming behind. By the 1930s, serious overgrazing and a lingering drought had devastated most Western grazing lands.

In 1934, the government stepped in again and passed the *Taylor Grazing Act*. The new law created the Division of Grazing and they began to issue permits for grazing rights on public lands. In 1946, the General Land Office merged with Grazing Service to create the Bureau of Land Management (BLM.) The grazing act led to the formation of a local committee to help address the problems of the driveway and one of the key issues was water.

In 1935, the newly formed Civilian Conservation Corps (CCC) came to help with the problems on the trail, and water was one of the first problems they addressed. They drilled wells every ten miles along the driveway, and installed windmills and stock tanks. This was considered to be the average distance the cattle moved in a day and it was the two-day average for sheep. Another team set the boundaries and built the fences. After these improvements were completed, it was said that the cattle were fatter by the time they reached Magdalena than they were when they left Springerville.

This lonesome corner of the West was used for centuries by Native Americans. Signs of prehistoric bison as well as Folsom man more than 10,000 years old have been found in the region. The area was also well-known to the U.S. Army for their long running battles with the Apache and Navaho Indians.

The new stockyards at Magdalena shipped their first herd of cattle in 1885 and its last herd in 1970, officially closing out eighty-five years of driveway history.

In 2008, Datil, New Mexico rancher Dave Farr was interviewed by the Bureau of Land Management. He brought in the last herd of cattle to ever use the driveway and provided a great account of the drives and an unbelievable collection of period photographs. The interviewer asked Farr how many horses he took on a drive, "Not many

— about three each. "You'd ride one in the mornin', and one in the afternoon and one at night, and have an extra horse."

Talking about one of the chuckwagon cooks, Farr noted, "He'd fry eggs in the mornin' in the dutch oven. And he'd use a big tablespoon to get the grease over them eggs and then he'd fish the egg out and put it on your plate with a lot of grease, and sometimes ask you — do you want an extra spoon of grease?"

On the closing of the driveway, Farr was ready to concede to the trucking industry. "Well, my father woulda preferred drivin' them to Magdalena, but we didn't. We didn't moan and groan over it, it's really kinda easier to round up the cattle, and load 'em on a truck and you're done."

For 85 years the driveway saw hundreds of thousands of cattle and sheep make their way from the wide-open country to the railhead at Magdalena and on to the markets. Today's modern highway 60 covers the basic route of the old driveway from Springerville, Arizona to Magdalena, New Mexico. Along the way are historic markers and campgrounds marking the trail of the longest operating livestock driveway in the country — make a drive yourself and see what it's like.

Mesa Verde's Cattle

A Cliff House in the Ute Mountain Park © Bert Entwistle

In the extreme southwest corner of Colorado, lies some of the most remote, desolate, and strikingly beautiful landscapes in America. Standing tall above the surrounding countryside is one of the largest mesas in the west, *Mesa Verde,* or Green Table in English. It climbs from 6,000' to more than 8,500' in elevation and contains more treasures of the ancient people and their world than could be seen or explored in a long lifetime.

Some of the first white men to see and explore this fantastic place were the brothers of the Wetherill family. After moving to Mancos, Colorado from Kansas in 1881, their father, Benjamin Kite Wetherill, and his wife Marion, established a ranch about eight miles southwest of town along the Mancos River and settled into the life of a pioneering cattle operation. After building two cabins of pine logs sixteen-feet square with twelve-feet between them (designed to be closed in at a later date), they built their fireplace with river cobblestones and logs and dirt for the roof. When they finished the cabins, they dug a long irrigation canal from the river to the front of the house. When the canal was done, they planted Cottonwood trees cut from the river bank around the property and named their new home the *Alamo Ranch*, or Cottonwood Ranch in English. In time, they developed a good herd of cattle and grew more than enough food on their irrigated ground. Along with their milk cows and chickens, wild game like deer, turkey, grouse, and fish were plentiful, and berries in season kept them well fed.

Benjamin Alfred (Al) Wetherill was the second son of Benjamin Kite, and the one he always said had, "a greater bump of curiosity than the rest of the boys." He kept a journal, and eventually included the story of his early days as a cowboy in his autobiography.

"Until cowboys have a good bunch of cattle, they are called cowboys; after that, he is always a cowman," wrote Wetherill. *"And in the early days, the makings of a cowman was to own a good horse, a long rope and a branding iron."*

Like most operations of the time, the Wetherill's ran their cattle in the canyons, meadows, and tops of the mesas whenever they could. The reality of the situation was that the *Mesa Verde* itself was part of the Southern Ute Reservation, and conflicts with their Indian neighbors would prove to be inevitable. One day, Al rode up on a cattleman and a Ute looking at a fresh brand on a cow. After a moment:

"The Indian drew his six-shooter and shot his gun, and the cow fell dead right where she stood. Then the Indian said — my cow, my grass."

Any potential ownership dispute was ended then and there. In time Al wrote:

"The cattle became like wild animals, and it was necessary to hunt constantly and look everywhere for strays so as not to overlook the hiding places of a wandering bunch of the Pinon splitters."

By 1887, the Utes had given the Wetherill's permission to run their cattle on the reservation land, and most of the conflicts ended, but the cattle still wandered into the farthest reaches of the roughest terrain.

By the time the Wetherill's built their ranch, it was already well known that the area had ruins from a much older civilization, but the deepest interior areas were little known to the outside world until the cowboys and cowmen of the day began searching for better pasture and missing cattle. The Wetherill's saw early ruins nearly everywhere they went, and in 1882 Al discovered the large ruin that would become known as the *Sandal House.* Whenever they found fresh ruins, they made time to explore and begin to document what they found. They scheduled trips into the area just to explore, and began to bring in outsiders that wanted to see for themselves and invited people from the scientific community to examine the ruins.

In 1885, Al first viewed the largest and most famous ruin, *Cliff House,* from the bottom of the canon. In 1888, brother Richard Wetherill and cowboy Charles Mason rediscovered *Cliff House* from the top of the adjacent rimrock. The Wetherill's thought that calling the area a mesa was a little misleading. "It is cut up with such a labyrinth of canons," noted Al, "that it consists entirely of canons with just a few ridges of soil to hold them together."

In time, the brothers began to devote more time to exploring the ruins and less time to the ranch. They began to take sightseers and archaeologists into the canons. As they discovered new sites, they cleaned out the ruins and built a museum at the Alamo Ranch, carefully documenting their finds and putting them on display. Tourists flocked to Mancos to see the famous museum and take guided trips to the sites, and the Wetherill family provided a register so they would know who traveled the farthest to see the ruins.

When the world got familiar with the site, it brought hoards of pothunters tearing into the ruins taking everything they could carry. Even the government was involved in looting the ruins. Al Wetherill considered the government and the pothunters to be equally as bad as far as looting the sites went.

In 1906, President Teddy Roosevelt created *Mesa Verde National Park,* in his words, "To Preserve the Works of Man." In 1966, it was listed on the *National Register of Historic Places,* and in 1978 was designated a *World Heritage Site.* The park consists of more than 50,000 acres with more than 4,000 ruins in all stages of stabilization and restoration. To the southeast, is the *Ute Mountain Tribal Park,* covering more than 120,000 acres with thousands more ruins. This park is private land and owned by Utes. To tour the park you must make arrangements with Indian management in Towaoc, Colorado. National Geographic Traveler Magazine recently named it one of the 80 top destinations in the world to visit.

Time has not always been good to the Wetherill legacy. They went from discoverer of many of the greatest ruins, to being called pot hunters and looters, and finally, after Al Wetherill's records and journals were published to the world and many historians stepped up to defend their legacy, they now have the recognition they deserve. The Wetherill brothers are rightly considered to be the first white men to see many of the greatest ruins of all times — in many cases while looking for the their wild, wandering cattle.

One Man's Legacy

"The Silk Robe" C.M. Russell - 1890

Tucked away in the southeastern corner of Colorado is Prowers County. To the west is Bent County, to the south is Baca County and the Oklahoma Panhandle. The east side shares its border with Kansas. It's a remote place and lonesome to those who don't love it — flat, wide open, and prone to wild swings in the weather. The history of this desolate country plays like a panorama of the opening of the American West. Buffalo, Indians, wagon trains, and soldiers fill the pages of local history books. Dinosaurs and prehistoric man claimed this corner long before modern man.

In 1856, at age 18, John Wesley Prowers, from Jackson County, Missouri, connected with a man named Robert Miller, the Indian agent for the Upper Arkansas River Indian Agency. They hauled a wagon train full of goods for delivery to William Bent's New Fort. The frontier outpost sat next to the Arkansas River, alongside the Santa Fe Trail.

Miller's agency distributed the annual annuity goods to the Apaches, Comanches, Kiowas, Arapahos, and Cheyennes; tribes that were signed to the recent Treaty of Fort Wise. Prowers became the clerk for Col. William Bent, owner of the fort and an Indian trader in the area since 1826. Taking a job with Bent, he was in charge of making at least ten wagon trips across the prairie to supply goods to the remote fort. For seven years he travelled the plains for Bent, hauling freight and trading with the Indian tribes for skins and buffalo robes. Prowers made several trips to Fort Union in the New Mexico Territory, and one to Fort Laramie in the Wyoming Territory. He personally commanded twenty-two different round trips across the prairie.

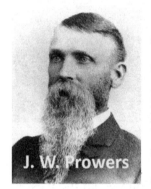

In 1861, John Prowers married fifteen-year-old Amache Ochinee, daughter of Cheyenne Chief Ochinee, also known as One-Eyed Chief. By 1858, the discovery of gold in the Rockies had brought waves of miners looking for their fortunes. Indian land and existing treaties meant little to those looking for gold and a new life. The Indians, angered by the depredations of the whites, disavowed the terms of the treaty made by their chiefs and continued to live by the old ways. Deadly conflicts between them and the new whites became common. As friends and advocates for all the tribes, John and Amache worked hard to assure fair treatment and keep the relations stable.

On November 11, 1864, a band of Cheyennes and Arapahos, mostly women, children, and men too old to fight, were camped at a place called Sand Creek. Chief Black Kettle of the Cheyenne and Prowers father-in-law, One-Eyed-Chief, were sleeping peacefully when they were awakened by rifle shots and the sound of charging horses.

The Colorado Militia, formed by territorial governor John Evans and led by Col. John Chivington, had adopted a take no prisoners attitude when confronting Indians. Guided from the fort by famed mountain man and trapper Jim Beckworth, they reached a ridge overlooking the encampment. Chivington, anxious to engage in the "battle," gave the order to attack as soon as it was light enough to see. 700 soldiers attacked the camp in full fury killing approximately 160 women, children, and old men, wounding many more. No adult warriors were present at the camp.

Kit Carson, close friend of both One-Eyed-Chief and John and Amache Prowers, expressed public outrage. Chivington publically declared a great victory telling the committee investigating the incident they had killed an estimated 500-600 warriors. Only four soldiers were killed and 21 wounded in this "Sand Creek Battle." Col. Chivington, a Methodist minister before his military service, was stripped of his command but never prosecuted.

"Damn any man who sympathizes with the Indians! . . . I have come to kill Indians, and believe it is right and honorable to use any means under God's heaven to kill Indians. . . . Kill and scalp all, big and little; nits make lice."
Col. John Milton Chivington

In 1868, Prowers and his wife, tired of the freight business, bought a farm near Boggsville, the temporary seat of Bent County. Turning his efforts to cattle ranching, he began to grow his ranch and paid close attention to the genetics of his herd. He introduced the Hereford breed to the area, and in 1871, bought a famous bull called *Gentle the Twelfth* from a breeder out of Canada.

In the next decade, he increased his holdings to 80,000 acres of fenced land including forty miles of river frontage on both sides of the Arkansas River. Each surviving family member of Sand Creek Massacre received a section of land by the government as an attempt to make reparations to the survivors of the massacre. In this area were the original 640-acre tracts given to Amache Ochinee, her mother, and two of the Prowers daughters. It became the foundation of the Prowers ranch. In time, the ranch also controlled nearly 400,000 acres of open range. Within twenty years he owned as many as 70,000 head of well bred cattle carrying the Box B and the Bar X brands. He eventually built his own slaughter house in Las Animas to cut out expenses and sold beef and live cattle.

Prowers was appointed Bent County Commissioner by the governor, and in 1873 he was chosen to represent the county in the State House as an Independent candidate. He served on numerous agricultural committees including irrigation and stock laws

Prowers believed that the government should lease the thousands of acres of open range it held and collect revenue from it. Never one to sit back and reflect, he had interests in water and wildlife and was one of the first in the area to introduce irrigation to his ranch. He dug miles of ditches on his property to water his crops and his stock. He introduced whitetail deer to the area near the mouth of the Purgatoire River and brought in Bob White quail and prairie chickens.

John and Amache had nine children, eight living to adulthood. They sent all of them, boys and girls both, to college. In 1889, five years after his death, Colorado split Bent County into two counties. The western half was named Prowers County in his honor.

Today, the site of the Sand Creek Massacre, in Prowers County, is a designated historic site managed by the National Park Service. Thirty-five miles from Sand Creek is a more modern historic site — the Granada War Relocation Center. After the bombing of Pearl Harbor, President Roosevelt signed Executive Order 9066 incarcerating more than 7,000 Japanese Americans in the prairie camp in Prowers County. The camp was surrounded by barbed wire fences and machine gun towers. Soon the camp became known as *Camp Amache*. A little farther west is Bents frontier fort, rebuilt to original specifications. Look closely when you drive through Prowers County and you can feel the real history of the American West pulling you in.

Painted History

Painting of Auroch from a cave in France about 15,000 Years-Old

Great artists are a truly unique group of people, they seem to have something inside of them that most of us don't. It's as though they feel the need to get it out for all the world to see, or perhaps for their own pleasure, but they continue to make their art regardless of the motivation. Their art can be expressed in many different ways. Some can turn a plain block of stone into a statue that is so beautiful it can take your breath away, and some can bring the stories of the bible alive on the ceiling of a tiny Roman chapel. But tens-of-thousands of years before those fantastic works were created, the artist worked in a much more modest medium — the walls of a cave.

Somewhere around 30,000 years ago, a group of early hunter-gatherers used a cave in what is now Southern France for shelter. Scientific evidence shows that many groups of people used that same cave as a home base over thousands of years. What those early people left for us is one of the most extraordinary art galleries in the world. The cave, called the *Chauvet-Pont-d'-Arc* had been sealed by a landslide thousands of years ago, and just rediscovered in 1994.

For whatever reason, the people of the day decided to record what was important to them at the moment, on the walls of their cave. It may have been for their religion, or it may just have been the hunters recording what they saw outside the cave every day. It's also possible it was just a little male chest thumping to impress the females of the clan. The ones who cleaned off the rock and painted these remarkable scenes were truly artists with a need to record their life. The paintings of animals like bears, rhinoceroses, tigers, and buffalos give a vivid look at the world 30,000 years ago. The art on these walls, and in other locations, were not crude, rough impressions, but beautiful, flowing paintings that are as fine as any contemporary work.

Among the paintings were images of the *aurochs,* a wild cow, the ancient relative of our modern-day breeds. This early version of our modern bovine was a giant, scary beast, feared by the hunters, but prized for their meat and hides. A mature aurochs bull measured as much as six and a half feet at the shoulder and weighed in at 2,000 pounds or more. With enormous forward looking horns, they were black, long legged, and wildly dangerous. The aurochs ranged from Europe to China and India. Discovered in the ruins of the Catalhoyuk historical site in Turkey, was art depicting a giant aurochs doing battle with dozens of hunters. From the painting, it appears the hunters are getting the worst of it.

Physically, these ancient bovines resembled a Spanish fighting bull but were much larger. Earliest domestication of the aurochs is recorded around 10,000-8,000 years ago, possibly somewhere around the Indian subcontinent. Generally speaking, domestication means human control over breeding. As these ancient animals began the long path to the breeds we have today, society, now with a steady source of protein from cattle and other animals, became less and less dependent on hunting. They began to live in larger groups and experimented with other forms of agriculture. DNA testing suggests that today's domestic breeds originated in Iran from about eighty animals over 10,000 years ago.

After the aurochs were domesticated, people began to breed them for different traits, and in time, the pure aurochs' line was no longer needed as a resource. A few wealthy landowners, most notably in the Balkan countries, kept a few around in their pastures. Julius Caesar wrote about the animal in his account of the Gallic War:

"These are a little below the elephant in size, and of the appearance, color and shape of a bull. Their strength and speed are extraordinary; they spare neither man or wild beast which they have espied."

The last known pure aurochs died in Poland in 1627. There has always been an aura of mystery around the wild aurochs. Today, it's still used as a symbol for different countries and states throughout Europe. Vikings often used the image on their shields and decorations. In the days of the nomadic hunter-gatherer clans, to kill an aurochs and display the horns was a great feat and a display of the hunter's courage. Drinking cups made from the horns of the aurochs and embellished with silver were common trophies.

It's a testament to the power and importance of this wild creature, that its image is found in archaeological sites throughout Europe, Asia, and India. Today, its ancestors now feed a large share of the world's population.

The idea of bringing back the aurochs has always been popular in Europe, and in recent years efforts to bring it back have been generating a lot of interest. Wildlife conservation groups feel that the lack of large grazing animals is detrimental to the large open areas. In the nineteen-twenties and thirties, a pair of German brothers by the name of Heck tried to back breed the aurochs. In the end they developed a breed called Heck cattle. These animals didn't resemble the aurochs in size, appearance, or behavior.

A Dutch organization called the *Taurus Foundation,* working with the *European Wildlife Foundation*, believes that if they can reintroduce a grazing animal like the aurochs, they will help manage these open landscapes and in turn help the other species in the area. The aurochs would be brought back by careful cross-breeding from the most primitive cattle with the closest traits. Genetic scientists believe that by 2025, the herd will be as close as possible to the original aurochs. The new breed will be called the *Taurus.*

Even 30,000 years ago the people who produced these magnificent cave drawings were catering to that need to express themselves. Fortunately, for all of us, these ancient artists, using the walls of their cave as their canvas and readily available materials like charcoal and ochre, left their vision of what life was like 30,000 years ago when giant cattle, *"a little below the elephant in size"* roamed the countryside.

The Queen's Calf

Engraving of Brown's Park at the Gates of Ladore

The feud with the old two-bar outfit began when Anne Bassett was just eight years old. It all started over an abandoned calf she found along the Green River that she named *Dixie Burr*. In the spring of 1883, when she found the lost 'dogie', life in Brown's Hole, or Brown's Park, as the locals called it, was still as remote as any place in the country. This wild corner of Colorado had been home to local Indians for hundreds of years. In time, mountain men, outlaws, settlers and ranchers claimed this spectacular valley for themselves.

After Anne's family moved into the park, her father started a modest cattle operation. Working with ranchers and buyers from Colorado, Utah, and Wyoming, the family flourished in the lush, well watered valley along the river. As the locals raised their families and their livestock, they created a thriving sense of community in the park. Soon, the big cattle outfits of the day took notice and began to flood large herds into the area to take advantage of the open range. Anne spoke of this in her autobiography, *Beef on the Hoof!*

"Vast, northward moving herds from Texas took over all the range in Wyoming and were on the march to Colorado. On they came relentlessly, that moving sea of hides and horns, devouring and spreading like a gigantic flood."

Anne found the fresh calf camped out in the Bassett's pasture, left over from a drive that passed close to their ranch. She knew it needed milk, so she moved her to the barn and told her mother about it. Her mother's advice was not what she hoped for "She immediately made it clear to me that I could feed and care for the calf, but as soon as it could eat grass, I must turn it on the range, for I knew very well that it belonged to Mister Fisher (it carried the brand and ear mark of Fisher's Middlesex Cattle Co.—the two-bar brand)."

Anne Bassett

The Middlesex office was in Rock Springs, a hundred miles away, and Anne talked her father into taking her along on his next trip for supplies, a ten-day ride from home. She intended to make a deal with Fisher for the calf. After explaining to him how she came onto the calf, her father was more than shocked to hear his little girl offer to trade one of his registered yearlings for the scrawny little longhorn. Fisher told her that he would not accept such an uneven trade, but that he would make a gift of her instead, saying that the calf would have died anyway. Anne finally had her calf and that was that — or so she thought.

After nursing the calf back to health, she turned it out with the other cattle as a yearling. During that time, Fisher had resigned from the Middlesex Company and the new manager was informed of the Brown's Park calf that carried the two-bar brand. There were enough cowboys working in the park from the different outfits, that it was common for them to spot a missing animal with their brand and round it up with the rest of the cows they were moving.

When Anne realized that Dixie Burr was missing from the pasture, she headed straight for the herd that just passed the ranch. When she caught up with the two-bar outfit, she searched out Dixie Burr and tried to take her back home. Suddenly, the cowboys found themselves faced with a young girl, dressed in buckskins, screaming at them for stealing her calf. Anne went at the cowboy like a demon possessed.

"I began whipping him over the head with my quirt. My slashing him over the head and face turned him plum sour, and he took on the work of properly educating and chastising me"

Finally, one of the cowboys, named Joe Martin, sided with Anne and against the cow boss, a man named Roark, and the fight was on. When all was said and done, Roark was soundly beaten and Anne took the calf home and didn't tell the family about the confrontation. She knew, however, that she would have to do something about the brand so it didn't happen again.

"I took her to an out of the way place and tied her up tight. I built a fire and put a branding ring in it. I made the two-bars into a pig-pen brand by adding two more lines at right angles to the bars."

Roark, suffering from the embarrassment of the beating, began to snoop around for the calf, and finally found her, now showing an obviously altered brand. He rode directly for the sheriff's office in Hahn's Peak, over a hundred miles away.

"He swore out warrants for nearly everyone in Brown's Park except father. He included men and women alike."

When the people of Brown's Park got their warrants, they didn't really know anything about the calf, but did their duty and appeared as directed. The case was immediately dismissed for lack of evidence. Roark lost his case and damaged his reputation even more, all over a little girl and a lost calf.

"As a result of my child efforts to protect a cherished pet from brutality," wrote Anne, *"Brown's Park was branded as a home for rustlers, and the lying rumor was widely circulated that no good can come from Brown's Park."*

Anne was formally educated in a good Eastern boarding school, and returned a well-educated and beautiful woman, and picked up her feud right where she left off — harassing the two-bar and every other big outfit in the area. She was said to have a habit of running their cattle off cliffs and into the river, and rustling their stock whenever she could. Nicknamed "Queen Anne," she continued to stand up for the small operations and make life hell for the big ones. The range war in Brown's Park was nearly as bad as the more famous one in Johnson County, Wyoming. Big cattlemen brought in killers like Tom Horn, to try and drive the locals out. Queen Anne once said that even though she was acquitted in court, she did everything they said she did, *"and a whole lot more . . ."* In the end, the skinny, muddy, burr covered calf she found that day did more for the rights of settlers than the killers sent from the big cattle operations ever did to hurt them.

The Genuine Real McCoy

Engraving of Joe McCoy's Stockyard and Hotel - 1880

"*Hell is now in session in Abilene . . .*" trumpeted the July 1867 headline from the *Topeka Commonwealth* newspaper. With the opening of Joe McCoy's Great Western Stockyards, the tiny settlement in Central Kansas had turned into the wildest town in the West almost overnight.

McCoy, a cattle buyer originally from Illinois, had seen states like Missouri shut down the huge trail drives from Texas to Kansas City and other locations. The local farmers and ranchers were beginning to block access to their ground. Longhorns driven up from the South carried ticks that spread a disease called Texas Fever or Spanish Fever, depending who you wanted to blame. The Longhorns were the carriers, but were generally immune to it, but the local cattle were devastated by the disease. The ticks left behind by the large herds killed thousands of animals not resistant to them.

The Texas ranchers and McCoy were making more money than ever. In 1866, the Texas market was paying about four-dollars per head compared to a more lucrative forty-dollars a head in the North. McCoy knew that the Union Pacific railroad was interested in the possibility of expanding their rails to places like Abilene, and he lobbied them hard for a new line, promising to fill their cars with prime, Texas beef. He then set to work plotting a course for the new Abilene Trail, a route that would connect to the North end of the Chisholm Trail.

While marking the route and looking for watering stops along the way, he founded the village of Newton, Kansas. Once the railroad committed to a new line, he made a 600-acre purchase outside of Abilene for his stock yard and headed for Texas to advertise the new venture. After investing $5,000 in flyers and advertising, he headed back home to prepare for the rush of business. The first drives of Texas cattle arrived in Abilene in August, and the first carloads were shipped to Chicago in September. The new system was a dramatic success and by 1870, the new Kansas cow town elected McCoy for their mayor. It didn't take long before he came to realize that with the success of the cattle business came everything that most towns didn't want — gambling, prostitution, gunmen, and murder. Different cattle drives often ended at nearly the same time, and as many as 5,000 cowboys, their pockets flush with trail drive pay, all landed in Abilene at the same time looking for excitement.

The wild little cow town was run by the brothel and saloon owners, and the mayor brought in one new sheriff after another. Some tried a no guns in city limits law and failed miserably. After a few days, some of the newly appointed lawmen simply disappeared, not wanting any part of Abilene's problems. In 1870, a fearless new marshal, named Tom "Bear River" Smith, a former professional boxer, started out well but died in a murderous ambush by two locals. The killers were caught and sent to prison for life, but the job of sheriff was open again. In frustration, Mayor McCoy began the search for someone to stand up to the lawless crowd. In April of 1871, he found the man for the job. James Butler Hickok, aka Wild Bill, became the next sheriff of Abilene. Hickok's fearsome reputation served him well when he came to town.

The very sight of him walking down the street with his wide-brimmed hat, long flowing hair and silver-plated .36 caliber Navy Colts tucked into his belt commanded more than a little attention from the local gunmen. Mayor McCoy had hired him for $150 a month plus a percentage of the fines he collected.

Wild Bill did a good job of keeping the cowboys in line until a saloon owner named Phil Coe, decided it was time to settle an old dispute. While working a street brawl, Hickok ordered him arrested for firing his weapon in the city. Suddenly Coe turned his pistol on Hickok, however, before he could fire, Bill shot him twice in the chest killing him instantly.

In the chaos of the moment, Hickok caught a fleeting glimpse of a man running toward him and fired two quick shots, hitting and killing his friend, Deputy Marshall Mike Williams, who was rushing to assist him. The event haunted Wild Bill for the rest of his life, and ended his career as a lawman.

By 1872, the railroads had pushed farther South and West, and the cattle business was shifting to the newer cow towns closer to Texas. As the cow business slowed, Abilene began to return to a quiet prairie town. After leaving the mayor's office in 1873, McCoy moved to Kansas City to pursue new business prospects. Kansas City had become the leading cattle market in the West. After a few years, he invested in the meat processing side of the cattle trade, and traveled extensively throughout the West and the Southwest to learn all aspects of the business.

In 1881, he was hired by the Cherokee Indians as an agent to collect all the revenue on lands owned by the Indians, and eventually moved to Oklahoma. He was then appointed to the job of Superintendant of the Range Cattle Dept. by the U.S. Census Bureau.

Joe McCoy lived his last years in Kansas City, passing away October 19, 1915. Later in life, he had authored a book entitled *Historic Sketches of the Cattle Trade of the West and Southwest*. Never was anyone better qualified to write about the early days of the cattle drives, the railroads, or the wild cow towns than the man who connected the live cattle drives to the railroads of Kansas than McCoy.

One of the more famous stories tells of McCoy's trip to Chicago for his meeting with the railroad officials. When he pitched his idea, he bragged that he would bring, "200,000 head of cattle to their market in ten years." In reality, he brought in nearly two million head over four years. The phrase "It's the real McCoy," is said to be coined by people who knew him as an early visionary and giant figure in the early Kansas cattle business — a great tribute to a legendary cattle man.

Boss of the Plains

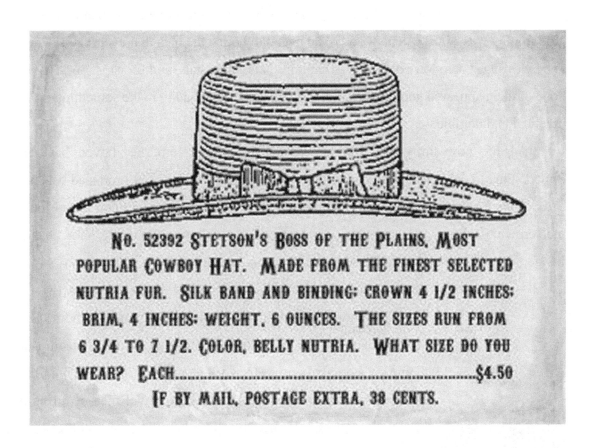

The No Name Hat Manufacturing Company had its start in downtown East Orange, New Jersey. Stephen Stetson was a long-time hatter said to have made hats for people like George Washington and his contemporaries.

The seventh of Stephen and Susanna Stetson's twelve children was a slight, rather sickly son named John Batterson Stetson. John worked with his father in the company until he was diagnosed as terminally ill with tuberculosis. Having been turned down for service in the war because of poor health, he took this as his cue to see the American West, something that had been on his mind for years. He was afraid this was the only chance he might have and within weeks of the diagnosis, he found himself in the Colorado goldfields, in the middle of the world of teamsters, settlers, mountain men, miners, and cowboys.

After trying his hand at gold mining, he saw his health improve dramatically, but his gold mining was a bust. As he worked, he began to notice the hats that everyone seemed to be wearing. There were derbies, sea captains hats, straw hats, and fur caps. To Stetson, none of them appeared good for frontier use. He decided he had to make something more practical and durable.

In 1865, now thirty-five and his gold-fever behind him, he moved back to Philadelphia to open his own hat company. The very first Stetson hat produced was the *Boss of the Plains* model (still in production today). The *Boss* had a plain round four-inch crown, flat brim, and was made of waterproof wool felt with a plain strap for a band. Inside, he put a sweatband embossed with his name in gold.

John B. Stetson

Stetson made a copy of this hat for every dealer that he met while out West and sent it to them with an order blank for more. The style caught on quickly and within a year the Stetson Hat Company started to grow. His working motto was, "None but That of Sterling Quality." In 1869 he had the novel idea of hiring a group of traveling salesmen to sell his hats. The sales of Stetson's hats were so great that he moved the factory and enlarged the operation again.

The company grew to be one of the largest hat manufacturers' in the world. They made the famous black U.S. Army campaign hat with the yellow braid around it that was used all over the West and is often used today in ceremonial occasions. They also supplied the iconic *Doughboy* hats made famous by our troops in WWI. Custer is said to have worn a Stetson into the Battle of the Little Bighorn.

As their popularity grew, the hat was dubbed a *Cowboy Hat* and it began to catch on with a part of society that didn't normally wear large hats, it became a fashion statement for many parts of society. Soon entertainers like Buffalo Bill were wearing them while performing. In time, the cowboy hat became a fixture in the movie business.

Stetson was ahead of his time with the process used to make the felt hat and his unique methods of sales and marketing. He was also a strong Baptist and believed in taking care of the people who worked for him. He believed that if he provided a good job, working conditions, and benefits, it would attract good workers. Stetson employees came to work every day to a clean, safe work place. The factory buildings were made of brick and had one of the first fire sprinkler systems ever installed. He insisted on nothing but the best tools, machinery, and materials.

He set up the most liberal apprentice program ever seen in the U.S. He paid above scale and offered bonuses attached to performance. He gave space in the factory for groups to use in various religious, social, and educational activities. He even organized a military company, had them trained, and set up an armory.

His concern for his workers' well being caused him to create a medical clinic that eventually grew into a full-scale hospital. For those unable to pay their medical bills, he provided a health program with the minimal charge of one dollar for every three months of service. If an employee could not afford it, they were cared for at no cost.

The hat factory grew to twenty-five buildings over nine acres. He built parks and housing for more than 5,000 employees during a time in history that was not known for good employee benefits.

Stetson's success made him a wealthy man, and near the end of his life, he had made the decision to give most of it away. Education was important to him and he built grammar schools and high schools. He helped to build Stetson and Temple Universities and donated so generously to DeLand University in Florida that they renamed it John B. Stetson University. In 1900, he created Stetson University Law School, the first law school in Florida.

The poor and homeless didn't go unnoticed to Stetson. In 1878 he helped create the Sunday Breakfast Rescue Mission, homeless shelter and soup kitchen in Philadelphia that is still in operation today. Stetson built a beautiful home in Philadelphia for the summer and another in DeLand, Florida for his family's winter residence. Next to the house, he built a schoolhouse to continue his children's education in the winter.

Stetson had a larger world view than did many of the men of his day. He saw wealth as something to share and something to help those that needed it. He led the way in everything he did and influenced a lot of his wealthy contemporaries. Stetson's close friends included such notables of the period as President Grover Cleveland, King Edward IV, the Mellons, the Astors, and the Vanderbilt's. Thomas Edison personally supervised the installation of electricity in the mansion, one of the first in the world to ever be designed and built with electricity from scratch.

John Batterson Stetson died in his DeLand mansion of a brain aneurism in 1906. His son George Henry continued to run the plant until 1971 when it was closed due to the decline in the hat business. In 1977, the family, true to Stetson's legacy, gave the buildings and land to the City of Philadelphia.

At its peak, the factory had 5,400 employees and sold 3,200,000 hats a year. Thousands of people prospered from one straightforward idea — a hat that protects you from the rain and snow and keeps the sun out of your eyes. The cowboy hat — the hat that won the hearts of America.

Theodore's Cattle

Theodore Roosevelt On His North Dakota Ranch

As Theodore stepped down from the train, he stood for a moment in the failing light taking it all in. It was early September of 1883, and his presence in the tiny Badlands town of Medora, was hardly enough to impress anyone. At five foot nine, painfully skinny, and the picture of poor health, his thin moustache and wire frame glasses made him a portrait of a dude, which he was. At twenty-five, he'd left the wealthy trappings of his New York City childhood to go out West and make his mark. He told his friends he

121

wanted to kill a wild buffalo before they were all gone. He was also there for his health, an issue he chose to keep private.

A Harvard graduate and politician by trade, he was woefully unprepared for the rigors of the Western lifestyle. He had really come to buy cattle, with plans to eventually start his own ranching business. Shortly after he arrived, he connected with three young Canadians, two brothers named Joe and Sylvane Ferris, and William Merrifield. Living on the Maltese Cross Ranch and grazing their cattle on public land, Roosevelt liked them immediately and the relationship between the four men lasted his entire lifetime. When he left Medora, he gave them a check for $14,000 to buy 450 head of cattle to start his operation.

His next order of business was to kill his buffalo and he enlisted Joe Ferris to guide him for the hunt. Everyone found out early that the little Eastern politician didn't like the name Teddy, but preferred his given name Theodore. Although early in the relationship some still called him Teddy, or four-eyes, or even sometimes worse names behind his back. Theodore Roosevelt Jr., the future President of the United States, had yet to prove himself to anyone in the Wild West, it wouldn't be long before the skeptics became true believers.

Ferris and Roosevelt scoured the Badlands on horseback for a week in all kinds of weather looking for a suitable trophy. After missing the first two that he shot at, he finally got his prize and ran up to the dead bison and let out a war whoop that could be heard for miles. After concluding his impromptu celebration, he handed Ferris a hundred-dollar bill and returned to Medora a happy man.

Back home in New York, his wife, Alice, was several months pregnant. The first letter she received from him was full of excitement and typical Roosevelt enthusiasm.

"Darling Wife, . . . by Jove, my usual bad luck in hunting has followed. I haven't killed anything, and afraid the hall will have to go without horns, for this trip at least. But I have had adventures enough at any rate."

Roosevelt talked about wagon breakdowns, missed shots at buffalo and antelope, rain, rough country, wind, and generally miserable conditions. He was sick from drinking bad water, and subsisted on dry crackers and rainwater for several days. He wrote that after spotting four bull buffalo,

". . . they crawled nearly a quarter mile on our bellies like snakes." By his own admission he hit the old bull *" . . . too far back, and the wound did not disable him."* The bull recovered enough to catch up with the others. *"The next hour was as exciting as any day I ever spent. . . . the bull turned to bay and charged me; the lunge of the formidable looking brute frightened my pony. He threw up his head and knocked the heavy rifle into my head."*

Roosevelt was half blinded from bad eyesight and blood in his eyes. The bull then charged Ferris and he escaped with his fast horse. Eventually, he claimed his trophy for the hallway, and after two weeks, it was time to return to New York to be with his wife when she gave birth.

Returning to Medora the following June was a bittersweet moment. His wife Alice had died in childbirth and his mother died on the same day in the same house. The pain was nearly more than Roosevelt could take, and he knew the wild loneliness of the Badlands country was the only thing that might save him from the depth of his depression. He once wrote that *"The Badlands look like Poe sounds."*

When he returned, he declared that ranching would forever be his occupation, bought more cattle and plunged into the business with his three Canadian partners and loved every moment of it.

This time he came to the Badlands prepared. He looked like the cover of one of the popular dime novels of the day. He was covered in buckskins, chaps, a giant hat, silver Conchos, and a big buckle. With his droopy mustache and glasses, he raised more than a few eyebrows. Many locals had a good laugh on him, but he didn't care, he was where he wanted to be, doing what he wanted to do.

Roosevelt wasn't the only one to take to the grasslands around Medora. A few months before Roosevelt arrived, a Frenchman with the impossible sounding name of Antoine Amedee-Marie-Vincent-Manca de Vallombrosa Marquis de Mores had also decided to stake his claim in the grass of the Badlands. Tall and handsome, he was also wealthy and arrogant. He decided to buy land rather than free range his stock. Soon he had 3,000 fenced acres and hundreds of cattle. Since there was now a rail line, he built a slaughter plant and planned to ship his meat east by rail. It was the Marquis de Mores who founded the town of Medora.

Between Roosevelt, the Marquis, and other area ranchers, the cattle business boomed. They could ship all the live cattle they wanted or butchered meat in the Marquis' new refrigerated railcars. It was a good time for Roosevelt. His childhood weaknesses and his asthma were long gone, he was now a healthy, even robust young man, toughened up by the country and the work. Theodore Roosevelt Jr. had been fashioned into the foundation of the Rough Rider he would become.

The legendary blizzard of '86-'87 changed their world and the cattle business quickly and forever. The ranchers of Wyoming, Montana, and the Dakota Badlands were wiped out. Every living animal on the open range was dead. In the spring, thawing rivers plugged up with the bloated carcasses of the cattlemen's futures. The Marquis had left early that winter, and after the storm, Roosevelt wrote that he had ". . . *rode for three days without seeing a live steer.*" The Badlands cattle business was done. As for the future president, he took the defeat in stride and moved on, but the Badlands of North Dakota had succeeded in shaping up a skinny, sickly rich kid into the man who now looks out proudly from Mount Rushmore overseeing what he helped build.

The Emperor of the Badlands

The Marquis de Mores Patrolling His North Dakota Ranch

Note: I wrote Theodore's Cattle and this one at different times, but they were both contemporaries of the same time and place so I put them together.

How exactly does one become an emperor in the Badlands of the Dakota Territory in 1883? For a strange little Frenchman, with the unlikely name of *Antoine-Amedee-Marie-Vincent-Manca Amat de Vallombrosa, Marquis de Mores et de Montemaggiore,* it took a large bankroll and an even larger than life persona to earn the title.

The Marquis was born in France in 1858, and at the age of eighteen claimed his rightful family title and inheritance. Joining the French military in 1879, he graduated from St. Cyr, the top military academy in the country. From there he was assigned to Saumur, the military cavalry school where he became an officer. While assigned to a post in Algiers, he had his first taste of battle and the first of many duels in his life.

Mores invested his inheritance in the stock market and other risky investments, and quickly lost everything. He had to go to his father in Cannes and ask for help, the Duke agreed to pay more than $100,000 to settle his debts.

While in Cannes, he met his future wife, *Medora von Hoffman*, at the von Hoffman winter home. After meeting her father, *Louis A. von Hoffman*, a wealthy New York banker, he began to court Medora. After his early failures left him broke, he found just what he had been looking for — a beautiful wife from a wealthy family. In 1882, they married, and after an extended honeymoon in Paris, moved to New York where the Marquis joined his father-in-law's wall street banking firm.

After pitching his father-in-law about his idea to transport dressed out beef in refrigerated railroad cars to the East, bypassing the giant Chicago packers, Louis agreed to finance his project if he could find suitable ground for the operation. With the last piece of his plan in place, he and his private secretary, William Van Driesche, travelled on the Northern Pacific to the tiny village called *Little Missouri*, where the rails crossed the Little Missouri River. After a thorough survey of the area, he chose a site near the river for the first step of his plan — the packing plant. The Marquis described the quality of the country in his journal:

> "These advantages are: . . . shortest haul to market, railroad facilities, water and ice to any extant, abundance of fuel in the shape of lignite, immense amounts of range, shortest haul to market, shelter and grass along the Little Missouri River."

The Marquis de Mores chose the place to start his business and named it Medora, a new town in honor of his new wife. He immediately found investors and formed the Northern Pacific Refrigerator Car Co. His plan was to slaughter 150 head a day, store them in ice houses scattered along the rail line, then ship them back east by refrigerated car. This would allow him to cut out the big Chicago plants like Swift and Armour and make more profit.

The slaughter house opened in October of 1883 and butchering began immediately. He began to buy more and more land under the provision of a government program called *Valentine Scrip*. The scrip was a document giving the owner the ability to buy unoccupied government land. He eventually owned thousands of acres of land from this program and from the purchase of railroad land.

Already a controversial figure around the badlands, the idea of a flamboyant, arrogant foreigner buying up so much of the prime land, caused a lot of resentment with the locals. Slim and strikingly handsome with a thin handlebar moustache, he rode the countryside with a Winchester rifle, Colt pistol, and custom-made buckskin clothes and horse tack. He rode the finest horses and built a large home that the locals dubbed the *Chateau de Mores*. Behind his back they called him the *Emperor of the Badlands*.

If he had any friends left after he bought the land, he lost the last of them when he fenced it off. This was the first time anyone had done anything like this and the locals were enraged. The emperor simply didn't care what anyone else thought, he was moving forward with his master plan.

Never one to stay still — or quiet, he forged ahead and built stores, houses, and a Catholic Church for himself and his family to have a place to worship. When a local named Frank O'Donnell got mad and cut his fence, Mores tracked him down and took a shot at him. He missed him but killed his horse. The locals, now madder than ever, dragged him to court and a jury acquitted him. It was not his first time in trouble with the law, he had been accused of the murder of a hunter named Riley Luffsey and acquitted.

While he was building his meat packing empire, he also started the Medora to Deadwood Stagecoach and Freight Co. in 1884. He claimed that he challenged Theodore Roosevelt to a duel in a letter and he refused. His freight business brought in a variety of food, including fish like salmon from as far away as the West coast. He was able to haul them cross-country in his refrigerated cars. Mores opened butcher shops in anticipation of new business, raised horses, sheep, and experimented with farming.

The longer he was in Medora, the more money his businesses lost and eventually his investors, including his father-in-law, pulled their funding. The Northern Pacific refused to give him the same low shipping rates as they gave to the Chicago Beef Trust, and Mores, already intolerant toward anyone who didn't agree with him, went into anti-Semitic rants claiming he was a victim of Jewish plots to ruin him, something that would influence him for the rest of his life.

To make it even worse, his beef was not selling as well as Chicago's because people didn't like the taste of beef raised on prairie grass. Chicago fattened their cattle on corn right at the stockyards. Drought and competition from other ranches added to his financial problem and by the fall of 1886 he knew the grand plan was dead. Hated by the locals and deeply in debt, the Emperor of the Badlands left forever. Back in France he had a 1.5 million-dollar debt, and again went to his father and father-in-law for help. Rejoining the military, he was sent to Africa. Hated by the Muslims, he was assassinated in 1896 by a Touareg fanatic.

The rein of the emperor lasted a short three years, but refrigerated train cars and trucks proved to be an idea that is still with us today.

The Lincoln County Cattle Queen

"When Guns Speak" C.M. Russell - 1898

Susan Hummer was born on a bitter cold December night in 1845 in the little town of Gettysburg, Pennsylvania. Her mother, Elizabeth, died when she was young and her father, Peter, remarried and moved the family to Eureka, Kansas. In 1873 she met and married attorney Alexander McSween from Atchison, Kansas. The pull of the open West was strong and in 1875 the couple made the long cross-country trip and settled in Lincoln, New Mexico. McSween found legal work and eventually took a job working for the Lawrence Murphy Company alongside a man named James Dolan. From that point forward, Susan McSween's life would never be the same, and certainly not anything she could have ever dreamed of.

The Lawrence Murphy Company had a firm grip on nearly every commercial business in Lincoln. McSween soon soured on Dolan and the company and went to work for wealthy English rancher, John Tunstall. Through their friendship, he became friends with John Chisum, a wealthy Texas cattleman now well established in the New Mexico Territory. In 1877 the three men formed their own business to rival the Murphy & Dolan Mercantile group and break the monopoly. They opened the J.H. Tunstall & Company, a mercantile and bank near their rivals.

The problems started immediately when an outlaw named Jesse Evans and two others working for Murphy and Dolan in what was called *The Santa Fe Ring*, shot and killed John Tunstall on February 18, 1878. This was the first killing of what was to become known to history as the *Lincoln County War*.

Susan McSween

In retaliation, McSween and Chisum gathered together a group of men called the *Regulators*, including Billy the Kid, to look out for their interests. The Regulators killed Sheriff William Brady, a strong Murphy-Dolan supporter, and several others to avenge the killing of Tunstall. The dust finally settled on the streets of Lincoln on July 19, 1878, when after five days of fighting, the U.S. Cavalry took control of the town and ended the fighting.

On the last day of the war, the McSween store and house were torched and as they attempted to flee, Alexander McSween and two others were shot and killed. Susan McSween witnessed the murder of her unarmed husband Alexander. In 1878, Territorial Governor, Lew Wallace, sent in Lincoln County Sheriff Pat Garrett to bring order back to the county. It took nearly three years for Garrett and his deputies to hunt down and kill the men involved, including Billy the Kid. The casualty estimates for both sides of the fight was approximately 22 men killed and 23 more wounded.

Seeking vengeance for her loss, the widow McSween hired a one-armed attorney named Huston Chapman, and tried to have the Murphy-Dolan faction arrested for murder, but Lincoln County declined to prosecute them. Then she had Chapman try and intercede with the governor on behalf of the Tunstall, McSween, Chisum Regulators. One of the last of the outlaws not yet captured or killed, Jesse Evans, shot and killed attorney Chapman and fled the territory, disappearing into history as the last of the Lincoln County killings.

In the aftermath of the war, McSween found herself near financial ruin and received help from the Tunstall family in England, serving as executer of John's estate. She settled her husband's estate and cleared her debts which left her a small amount to live on. In 1880, she met and married her second husband, George Barber, a surveyor for John Chisum.

Chisum gave her a gift of forty head of cattle and she began her own operation near Three Rivers. As the ranch grew larger, the marriage failed, and after the divorce, she took control of their 1,158 acre ranch. She laid out, built, organized, and ran the ranch herself, working under the Three Rivers Cattle Company brand. The experience in Lincoln County had taken its toll on her and she continued to fight for justice for Alexander and for respectability as a citizen and rancher.

The Three Rivers ranch became one of the largest in the territory and by 1890 was running as many as 5,000 head. In 1892, the *Old Abe Eagle* newspaper of Lincoln reported that she had driven 700 to 800 cattle to Engle, the most accessible railroad point, from which place they were shipped in *"38-foot New England Cars"* to the Jones and Nolan feed lots in Grand Summit and Strong City, Kansas. An article in the *New York Commercial Advertiser* stated:

"Near the town of White Oaks, New Mexico, lives one of the most remarkable women of this remarkable age, at the present time a visitor in this city (New York). The house in which she lives, a low, whitewashed adobe building, is covered with green vines and fitted out with rich carpets, artistic hangings, books and pictures, exquisite china and silver, and all the dainty belongings with which a refined woman wishes to surround

herself. The house was built with her own hands. The huge ranch on which it is located with its 8,000 cattle, is managed entirely by her. It is her that buys or takes up the land, selects and controls the men, buys, sells and transfers the cattle. She is also a skillful and intelligent prospector, and found the valuable silver mine on her territory."

By the middle of the 1890s, some estimates had reached as high as 8,000 head or more on the Three Rivers Ranch. The small silver mine she found on the property produced a steady supplement to the ranches bottom line. She became known as *The Cattle Queen* of the New Mexico Territory. Filling the ranch with fruit orchards from trees she was given by Chisum, she also put in grain fields for her stock, vegetable gardens, and berry patches.

In 1902, she sold her ranch and eventually settled into a cottage in White Oaks. Over the last twenty-seven-years of her life she lived quietly until the money she had saved was exhausted and she slowly sold off her jewelry and possessions to support herself. Her nephew, Edgar Shields, helped support her through her last years until she died of pneumonia on January 31, 1931, and he paid her funeral expenses. This larger than life western woman died penniless and unheralded in her small cottage. She is buried in the Cedarvale Cemetery in White Oaks. Sadly, the picture from the cemetery shows a small, unattended stone with her name misspelled as MacSween. It should, at the very least, read something like this:

***Susan McSween Barber**
1845 – 1931
A True American Pioneer

The Wild Corner

South East Arizona's No-Man's Land © Bert Entwistle

Traveling through Apache County, Arizona is a lot like taking a ride through the geology books of the American Southwest. It's a wide-open land, full of spectacular desert landscapes, deep canyons, wild grasslands, and mile after mile of extraordinary rock formations. The shapes and the colors of this beautiful place can capture your imagination and hold onto it for a lifetime. From the lowest desert cactus to the 11,400' Mount Baldy, Apache County is as diverse and beautiful as anywhere in the Southwest.

    The history of this region is just as wild and intense as the very land it was built on. A long, narrow county, it rubs up against New Mexico on the east and Utah on the north.

It also meets up with Colorado at the geographical intersection called Four Corners, where all four states come together at one point.

Inside its boundaries, three different Indian nations call it home — the Apache, the Navajo, and the Zuni. The Petrified Forest, Canyon de Chelly, and the Painted Desert, as well as a national forest and several wilderness areas add to the beauty of this remote part of the west. The southern half of the county is the most suitable for agriculture. The Little Colorado River gathers its headwaters here in the basin for its run to the Colorado River of the Grand Canyon. The south border is Arizona's famous Mogollon Rim.

The Mormons were the first Europeans to attempt permanent settlement in this remote part of the world. As early as 1847, they established a network of villages and farms along the Little Colorado that led to the founding of nearly two-dozen permanent towns. In 1879, the legislature split the area into several counties and again in 1902 split the far northeast part of the area into two counties, Navaho and Apache.

As civilization took hold, imported cattle were trailed into the area from West Texas and put on the native grassland of the Little Colorado River basin. By 1881, the railroad reached the area and the town of Holbrook became the center for all Eastern Arizona shipping. The beef speculators of the day were already gearing up for the next boom in the range cattle business.

In 1884, the Aztec Land & Cattle Company, founded by Edward Kinsley, set up business outside of Holbrook and quickly dominated the local market. The Atlantic & Pacific Railroad, in financial trouble at the time, sold Aztec 1,000,000 acres of their Arizona grassland for fifty-cents an acre. After buying several other outfits, they controlled nearly 2,000,000 acres at the height of their operation.

The first shipment to Aztec from Texas was 33,000 cattle and 2,200 horses. They adopted the *Hashknife* brand, an unusual design based on a common chuckwagon tool of the time. In the beginning, the small town embraced the business coming their way from the ranches, but grew tired of them in short order. Along with the wild cowboys came the usual drinking, gambling, and inevitable violence that plagued all booming cow towns of the time. Records show that in the year of 1886 there were twenty-six shooting deaths in Holbrook. Burton Mossman, the Aztec foreman, noted that most of them were Aztec cowboys at one time or another.

One settler in the area wrote: *"Thousands of longhorns ate the grass; riffraff and hell-hounds out of Texas ate the rancher's beef."* Another favorite target of the Hashknife cowboys was the sheep operations that were scattered all across the county. Although Aztec Land & Cattle was never officially a suspect, many of their cowboys were said to be involved in feuds and gunfights between the sheep men. The Pleasant Valley Sheep War was the deadliest conflict of its type in Arizona history.

After a few years of Aztecs practice of keeping everyone else from using their vast land holdings, the small farmers and ranchers were driven out by the cowboys of the Hashknife. The Aztec men herded thousands of sheep into the Little Colorado River and drowned them or shot up the herds.

On the morning of March 29, 1889, near the tiny town of Canyon Diablo, four cowboys in masks stopped a train and stole $1,500. The Railroad put up a five-hundred-dollar reward for the bandits and notified the Yavapai County Sheriff, Bucky O'Neill. He gathered together a quick posse and the chase was on. Recent snow made the men easy to track and after several days, they caught up with them just across the border in Utah.

For five more days they chased them, finally pinning them down in a remote Utah canyon and forcing them to surrender. The sheriff became a famous man all over the west, mostly due to his own publicity, and he told the *Tucson Star* that *"they were the worst desperadoes that ever operated in this western country."* In reality, they were just a couple of Hashknife cowboys looking for some excitement and a little extra cash.

By the turn of the century, the Aztec Land & Cattle Company was feeling the effects of their negative impact on the range and the people living in the area. They had flooded the range with Texas cattle they'd purchased at a low price. The low prices were offered because Texas had seriously overgrazed their own range with thousands of extra cattle on speculation. Without any thought to the Texan's dilemma, they began to overstock their range exactly as the Texas ranchers did — with the very same results.

After several years of failing business, the Aztec Land & Cattle Company gasped their last breath in 1902. Its history is full of wild cowboys, shootouts, sheep wars, and train robberies, but the biggest crime to this spectacular corner of the world is their legacy of range destruction and erosion caused by massive overgrazing. One early pioneer in the area wrote:

"When we came to Arizona in 1876, the hills and plains were covered with high grass and the country was not cut up with ravines and gullies as it is now. On the Little Colorado we could cut hay for miles and miles in every direction. The Aztec Cattle Company brought tens of thousands of cattle into the country, overstocked the range and fed out all the grass. Later, tens of thousands of cattle died because of drought and lack of feed and disease. The riverbanks were covered with dead carcasses."

Aztec Land & Cattle Company 1884 – 1902. *

The Queen City Throws a Party

First Stock show in Denver - 1909

In 1858, Denver, Colorado Territory was little more than a few coarse cabins and corrals set just above Cherry Creek where it joined the wandering South Platte River. Although the discovery of modest placer gold deposits in November of 1858 had sparked the earliest gold rush to the Colorado Rockies, the new town itself had very little to show for it.

That first winter of 1858-1859 the town had no more than 50 residents, many of them just waiting for spring so they could head to the mountains for newer gold fields or just about anywhere that showed a little more promise. In those early years, the city operated mainly as a hub for the plains agriculture business and supplier for the new Rocky Mountain mining districts springing up all over the mountains. However, carving out a place for Denver in the history of the West was proving to be a challenge for the new city fathers.

In April of 1859, the first newspaper in the region, *The Rocky Mountain News* was founded and the publisher, William N. Byers, wasted no time in aggressively editorializing his feelings about what was needed. Byers and the new Territorial Governor, John Evans, pushed a relentless campaign to remove all of the Indians from the territory. The Cheyenne and the Arapaho found themselves fighting for the life they had always known, and throughout the early seventies, the violence increased and living with the threat of attack became a way of life. As Byers and Evans saw it, the natives were just one more thing holding Denver back from becoming a first class city.

By the fall of 1867, Oliver Loving, of the Goodnight-Loving partnership brought the first large herd of cattle from Texas and New Mexico to the town of Denver. Though few locals realized it at the time, it was to become the beginning of a business that would help define the future for life in the Rocky Mountain West.

In 1870, Denver's world changed dramatically. Byers' never-ending push for progress now took up the cause for a railroad. When the Union Pacific finally hit Cheyenne, the Denver and Pacific Railroad was formed to serve the little town on the prairie. A former Territorial Governor, William Gilpin, a well-known and colorful orator of the day, was a champion of all things related to Denver and declared it a "*preeminently cosmopolitan*" city. Gilpin announced Denver as positioned for greatness as a natural stop between the Atlantic and the Pacific and the "commerce of mankind" was destined to flow through it.

When the first engine finally chugged its way into town from Cheyenne spitting cinders and belching stinking black smoke on June 24th 1870, the people of Denver loved it. To them it smelled like prosperity. The railroad became a like giant life preserver thrown to a floundering town and everyone grabbed on to it. For the next twenty years, the town grew from an 1870 census of 4,759 to more than 106,000 in 1890. Reporters and writers began to refer to her as the Queen City of the Plains or of the Rockies and soon she became known to many as simply the Queen City. The dream of Evans, Byers, Gilpin, and many others of the day was coming true, Denver was beginning to realize her potential at last.

By the end of the century, after two decades of steady growth, the Queen was beginning to feel the last days of her golden age slip away. The national depression of 1893 and the repeal of the Sherman Silver Purchase Act once again left her in the doldrums. As the civic leaders promoted economic diversity by pinning their future hopes on sugar beets, wheat, and manufacturing, an idea began to germinate in the minds of the cattlemen and movers of the day. Denver needed a show, a party some called it, to promote the very agriculture it has been touting to the world for years. The cattle business was a stable industry as the city grew for those first decades of life, why not have a stock show to show off a little? Although Denver had a meat packing industry and the Denver Livestock Exchange, it was modest compared to larger eastern rivals like Chicago and Kansas City.

The decision was made to open in January, along the banks of the South Platte River. January proved to be the perfect time for the stock growers and the new *Western Livestock Show* was born on January 29th 1906. For lack of a proper auditorium, the promoters erected a circus big top and *The Rocky Mountain News* reported that the local hotels were filling fast. The city declared "Denver Day at the Stock Show" and made January 31st a holiday and the citizens took full advantage of it attending by the thousands.

At the first show, a 1,150 lb. Shorthorn brought an unbelievable thirty-three cents a pound to its owner. Four different breeds were judged in that first show, Hereford's, Aberdeen-Angus, Shorthorn's and Galloway's and sirloin was ten cents a pound at the local vender's tent.

For the next hundred years, the Queen showed off her cattle, horses, and everything to do with Western agriculture. After a few years, the Queen was bestowed another title — *Cow Town of the Rockies*. By the twenties and thirties, the show was known worldwide as the place to be in January if you loved cattle and horses. Every breed of livestock was judged and bought and sold to people all over the country. As for the Queen, she couldn't be happier, her hotels and restaurants overflowed and prosperity, at least for two weeks in January, was back.

In 1931, the big party, by now called the *National Western Stock Show*, added its first pro rodeo and the party just kept getting better. Horse shows for every imaginable breed, Mexican rodeos, music, dancing horses and wild-west shows today provide a little taste of the Old West for everyone. Throughout the years, the Queen has gone through World Wars, depressions, oil and gas booms and busts, but the big party is still there, stronger than ever. The people of Denver and the Central Rockies have carried on a hundred-year love affair with their *Stock Show* and its ardor shows no sign of fading any time soon. Now, every year, hundreds of thousands of people come for the party and to pay their respects to the Queen and to mingle with those who helped make it the *Cow Town of the West*. At long last the Queen can take a break knowing her place in history is finally secure, and the party will go on.

U.S. Forest Service
110 Years and Counting

"We have fallen heirs to the most glorious heritage a people ever received, and each one must do his part if we wish to show that the nation is worthy of its good fortune."
__**Theodore Roosevelt**

Trappers Lake, Colorado, White River National Forest © Bert Entwistle

President Roosevelt was speaking about the natural beauty and the natural resources of America when he made that statement in 1905. The founding of the National Forest Service by Roosevelt gave the United States Department of Agriculture authority to bring new, modern methods to manage the national forests (previously referred to as "forest reserves") and grasslands. Influential men like explorers F.V. Hayden and John Wesley Powell, and photographer William Henry Jackson had been writing and speaking out about the need for preservation for years. The pressure by them and others of the period helped bring it to the American public. Photographs by Jackson and others proved to be a powerful tool, showing the world what the West was like.

Roosevelt's new head of the department, Gifford Pinchot, was born in Connecticut and a graduate of Yale. After college, he studied forestry in France. The new chief met with ranchers, farmers, loggers, and mining interests asking for their input. One of the first goals was preserving the watershed from serious erosion. Controlling the timber harvest in the forests and establishing controls on livestock grazing was an important first step.

As might be expected, there were dissenters on both sides of the issue. Some ranchers and timber companies argued that there should be no limits on trees or grass, that it should be "first come - first served." Many conservationists of the era, like John Muir, lobbied to have the public lands closed to all forms of extraction. They would have it designated as "foot or horseback only," more like a national park. Pinchot's response was that parks were to be preserved but forests were to be used. The same philosophy applied to grasslands. He felt it was better to allow "controlled grazing than unrestricted grazing or no grazing at all."

Pinchot said the key to the success of the new Forest Service was to maintain a stable and sustainable resource. The public lands would need to be scientifically managed to be successful. Unlimited exploitation of our natural resources was the mindset of the time. Convincing a crusty old rancher or timber man that they had to change their way of doing business was no easy job. Pinchot preached sustainability to everyone who would listen. By 1905, there was already severe overgrazing, streambed erosion, and problems at many mining claims. Hardrock and hydraulic mining each presented their own set of problems.

They would have to become involved in reforestation and rehabilitation of existing land damaged by fire. In August of 1910, the great fires of Western Montana and Eastern Idaho devastated three million acres of forest and cost seventy-two firefighters and thirteen civilians their lives. The fire raged unstoppably through the dry timber and even consumed their camps and destroyed what meager equipment they had. They did not have a store of firefighting supplies or nearly enough men to fight something so large.

The new National Forest Service now had another responsibility, writing new forest fire policy and building a specialized team to make it work. As they grew, they continued to take on more land and more responsibilities. They realized they not only needed to survey the boundaries of millions of acres of land all over the country, but at the same time provide access and trails into the many remote places they administered.

They not only issued cattle and sheep grazing permits but had to protect the forest from timber and grazing trespass, game poachers, and exploiters of all kinds. They also had the duty of watching for wildfires and began to build a system of fire lookouts throughout the heavily forested areas.

Until the formation of the USFS, grazing was free to the public. With the new organization came one of the more controversial management decisions, to charge for sheep and cattle grazing permits on the national forest land. Fees of twenty-five to thirty-five cents per head of cattle and horses were charged and less for sheep. The forest service set-up an office for grazing studies and researched the issue of carrying capacity to set the size of the allotments. In 1906 a law was passed to give ten percent of the receipts from the permits to the states to support public roads and schools. Two years later, that amount was raised to twenty-five percent.

The new position of National Forest Ranger was created and required both practical and written testing for the job. The applicant had to have extensive knowledge in subjects as diverse as livestock, forestry, surveying, and cabin construction. They had to be familiar with horses and be able to throw a diamond hitch on a pack horse or mule. They furnished their own horses and mules as well as their own rifles, pistols, and ammunition.

For the honor of becoming one of the new rangers, they received the sum of sixty dollars a month. These early days were sometimes referred to as the Stetson Hat era.

In 1919, Arthur H. Carhart was hired as the services first landscape architect. One of the earliest advocates of preserving particularly beautiful or scenic areas with wilderness designations, he explored the Boundary Waters Canoe Area in Minnesota and Trappers Lake, one of the most scenic areas in Colorado. Without his work and persistence, both would have been turned into private developments. Both were eventually made into National Wilderness Areas.

By the late thirties, the forest service began to develop new programs for fighting forest fires. In 1939 they developed the first smoke jumpers program. After a hundred and ten years in business, the U. S. National Forest Service now manages one hundred and fifty-four national forests and twenty national grasslands each with several ranger districts, some as large as a million acres. The lands are located in forty-four states, Puerto Rico and the Virgin Islands. These represent eight and a half percent of the United States land mass, about the size of Texas.

Some wilderness areas are inside the national forest system, such as the Flattops Wilderness Area in Colorado's White River National Forest, a treasure that was saved by visionary men like Carhart, Roosevelt, Pinchot, and Powell for use in our livelihood and recreation for our families for many generations to come.

The Texas Cattle Queen

"The Stampede" Fredric Remington

The romantic art of the American West generally shows a ruggedly handsome, well-weathered cowboy sitting on his equally good-looking horse. He's watching over a herd of cattle strung-out against the backdrop of a Charlie Russell painting. What it doesn't say is that not all the people riding those horses and watching the herd looked like the Marlboro Man — some of those riders were women. In most cases, their careers were as legitimate cattlewomen and ranchers. A few like Cattle Kate were not so legitimate and met her maker at the end of a Wyoming lynch rope.

Margaret Heffernan Borland was one of those women that was destined to prevail no matter what the world threw at her. The Heffernan family came to America as part of the first shipload of Irish colonists to arrive in coastal Texas in 1829. Two Irish entrepreneurs, John McGloin and John McMullen, knowing the political and economic problems in Ireland at the time, arranged to provide settlers for the early colonization of Texas. The offer of cheap land and a fresh opportunity was just what many of the dispossessed Irish citizens were looking for. John and Julia Heffernan, along with Margaret and her older sister Mary, settled along the coast near San Patricio, a rough, wild area a few miles north of Corpus Christie.

The timing for the new immigrants was perfect, the great cattle industry of the nineteenth century West was just getting started. Once they found their land, John began to establish himself as a rancher and farmer. Wild longhorn cattle of the time were gathered from the area between the Nueces and the Rio Grande Rivers, sold to the new homesteaders and the business of ranching in southeast Texas began to take hold.

Over the next few years, the family added two sons, John and James. Margaret's life took her first major setback when her father was killed in an Indian attack in 1836. Before the family could get over that, the Texas Revolution was in full swing and forced her mother to flee their home with the four kids. After the defeat of Santa Anna, they returned to their homestead and started over.

Margaret Borland

Margaret's marriage at age nineteen to a cowboy named Harrison Dunbar produced her first child, a daughter. Shortly after the birth, Dunbar lost a gunfight and was shot and killed in the streets of Victoria. For the next few years Margaret worked hard to keep herself and the baby fed and clothed until she met and married a local rancher named Milton Hardy. Her second marriage gave her two more daughters. Hardy died young in a cholera epidemic in 1855, leaving Margaret with three young daughters to care for.

Four years later she married her third husband, Alexander Borland, a well to do local rancher, and gave birth to three sons and one more daughter. Borland was a successful cattlewoman and highly respected in this part of Texas. It looked as though Margaret's string of bad luck had finally been broken. She became a partner in the ranch and worked with him for the next few years learning the business. Life on the ranch was good and the Borland family thrived. In 1867 the reality of living in the pioneer West reared its ugly head again. A yellow fever epidemic raged through the area claiming Alexander, four of the children, and an infant grandson.

Margaret was left as the owner of the ranch, now with more grief that anyone should ever have to deal with in a lifetime and three children that still needed raising. In time, she became a successful rancher in her own right. By 1873 she had bought and sold cattle, hired cowhands to do the work, and made investments in land, growing her herd to 10,000 head. She was said to be the largest cattle rancher in Texas at the time.

After freak blizzard of 1871-1872, thousands of cattle had perished. The buyers in the north were now paying $23 per head instead of the meager $9 offered in San Antonio. Margaret decided to drive 2,500 head north for 800 miles along the Chisholm Trail to Wichita, Kansas to take advantage of the higher prices.

In the spring of 1873, she mounted what may be the first trail drive ever led by a woman, and possibly the first by a woman in charge without a husband to accompany her. Choosing half-dozen trusted trail hands, she also took her two sons, still both under fifteen, her seven-year-old daughter, and a granddaughter only a few years old.

The drive began in Victoria, Texas, taking nearly three months to complete. When they reached the end of the trail, Margaret fell sick to what was generally referred to at the time as "trail fever" or "congestion of the brain." Margaret Heffernan Borland died in Wichita on July 5, 1873 after successfully moving 2,500 cattle up one of the most famous trails in history. She had lost her father and one husband to frontier violence, two husbands, one son, three daughters and one grandson to disease. She lived in a time and place that rarely celebrated the accomplishments of women. When she died at 49, she left a legacy of perseverance and hard work that was rarely, if ever, matched by men or women of any era.

The university of Texas in Austin has her collected files and papers from that period, and they document a fascinating time in the history of the early cattle industry. Receipts show payments for things like: $81.75 for wages and provisions for five hands to work six days gathering "beeves for shipment." In Wichita she purchased $7.00 in goods from Jake's Store. Another handwritten receipt shows $4.00 paid to the city sexton for "digging grave for son," and a receipt for $29.50 for the expenses of her last illness. Her remains were transported to Victoria for burial, a receipt for $119.00 for her coffin is in her papers.

Margaret's son-in-law, Victor Rose, a Civil War veteran and long-time Texas newspaper editor, wrote that she was:

". . . a woman of resolute will, and self-reliance, yet she was not one of the kindest mothers. She had, unaided, acquired a good education, her manners were ladylike, and when fortune smiled upon her at last in a pecuniary sense, she was perfectly at home in the drawing room of the cultured as if refinement had engrafted its polishing touches upon her in maiden-hood."

Margaret Borland left behind one daughter, a teacher, and two sons to run the ranch and carry on her incredible pioneer legacy.

The Willamette Cattle Company

"The Great Explorers" Fredric Remington

By 1837, Ewing Young had already lived an exciting and adventuresome life as a farmer along the Missouri River, a trader, trapper, and early pioneer of the Santa Fe Trail. In 1826, he formed a trapping party of more than a hundred men, exploring the Colorado River basin, New Mexico, Arizona, and Colorado. The large party was intended to discourage Indian attacks, but nobody told the Indians and violent encounters followed them throughout the trip. On his return to Taos, in 1827, Young and a partner set up a trading post selling goods supplied from St. Louis carried down the now popular Santa Fe Trail.

By 1830, Young was getting the itch to leave the relative comfort of Taos, and he formed another party that followed the Grand Canyon, through the Mojave Desert to the present-day San Fernando Valley, and on to the San Jaoquin River. Finding the area freshly trapped out, they moved on to the San Francisco Bay area where he re-supplied the party with goods from newly arrived ships. It was on this journey that he had his first encounter with Peter Skene Ogden, captain of the Hudson's Bay Company, the Canadian fur trapping giant. Though he didn't yet know it, the HBC would feature prominently in his future.

Ewing Young

Eventually, Young followed his lure for adventure North, exploring new trails from California to the Willamette Valley of Oregon while trailing a large herd of horses. Young was a gifted entrepreneur and went on to set up a grist mill, sawmill, and raise wheat as well as horses on his newly claimed ranch in the Chehalem Valley. His ranch soon became an economic center for the valley acting as a trading post, store, and even a bank.

By 1836, Young realized that the Willamette Valley was suffering from a stranglehold by the HBC, they controlled everything to do with certain parts of life in the Valley. The trapping company, working out of Fort Vancouver, had a monopoly on the cattle business and the locals were unable to start their own herds. In 1836, when Young was working on an idea for a distillery, he had a fortunate meeting with Lieutenant William A. Slacum. President Andrew Jackson had sent Slacum to the Oregon country to check out the economic and strategic affairs of the region and report back. Slacum was an American sailor and diplomat and possessed a presidential commission to check out the region.

Slacum eventually helped talk Young out of building a distillery and into a cattle-buying venture. Slacum and Young formed the Willamette Cattle Company to purchase cattle in California and drive them to Oregon. Slacum even provided $500.00 to get the business started. Young and Slacum decided to sail to California, buy the cattle, and take the herd North to the Oregon country.

Young found the cattle buying business tough in the San Jose area, but after several weeks of hard negotiating, Young and his American outfit finally put together a herd purchased for $3.00 a head from an assortment of Mexican generals and locals. With 729 head and another 40 horses, the herd turned North. It was the beginning of the end for the HBC monopoly in the Willamette Valley.

One of the American partners was Phillip Leget Edwards, a long-time trapper and explorer of the American West chosen as treasurer of the new company. Edwards kept what is thought to be history's first written account of an overland longhorn cattle drive. It was Edwards' writing that gives us a look at what life was really like in 1837, from the time the Willamette Cattle Company was formed to well into the drive to Oregon. On the purchase of cattle Edwards wrote:

"May 29th 1837, Mr. Young returned from San Solano, (San Francisco) having purchased seven hundred cattle at $3.00 a head, to be received 200 at a rancho of San Francisco and 500 at the mission of San Jose."

After many delays and a few dealings with the often-shady Generals, (Edwards reports paying a bribe of 1 rifle, 6 shirts and $20.00 cash to the local administrator) the herd was finally moved out. Young, Edwards and the rest of the company found their first big problem at the San Joaquin River — the cattle balked and broke and the fight to recover them was on. Edwards reports that the boiling sun, as well as the never relenting mosquitoes, added to the battle and they often resorted to using "skin lassos" from canoes and horses to try to capture their charges. The packhorse carrying their powder threw off her load in a pond and ruined it.

The crossing was a sign of things to come on the trip, and the Edwards journal entry of July 20t^{h} gives a sense of the difficulty.

"Little sleep, much fatigue! Hardly time to eat, many times!"Cattle breaking like so many evil spirits, and scattering to the four winds! Men, ill-natured and quarreling, growling

and cursing! Have, however, recovered the greater part of the lost cattle and purchased others. Another month like the last, God avert! Who can describe it?"

By mid-October, Young and crew reached the Willamette Valley after more than four months on the trail. After fighting Indians, heat, swollen rivers, and unfamiliar terrain, they arrived with 630 head in their herd. The cattle were estimated at $8.50 per head, with Young's share being 135 animals, setting him up as the largest rancher in the Oregon country. The impact of this historic drive was to loosen the iron grip the British had in the valley, (the HBC) and eventually led to opening up the future of the Willamette Valley.

Ewing Young's remarkable role in the history of the West and the settling of the Oregon country continued even after his death in 1841. Friends and neighbors found that Young died without a will and had no way to dispose of his estate. A temporary probate government was chosen to dispose of his property in a fair way. This probate government became the first formal authority in the region and the first step on the trail to statehood for Oregon. By 1843 the residents of Oregon formed their own provincial government, much to the dismay of the HBC, and sixteen years later became the 33rd state.

Although only 42 years old at his death, Young's path through history took him to most of the wildest parts of Western America. He left his mark on St. Louis, the Santa Fe Trail, the Grand Canyon, and California while under Mexican control and on to the Pacific Northwest. In the process, he headed up one of the most historic longhorn cattle drives in history and helped lay the foundations for agricultural life in Oregon. Today, Young's grave lies under a big Oak Tree overlooking his beloved Valley, the end of the road for many thousands of Oregon Trail immigrants who came after him.

The Wiregrass Herd

"Galloping Horseman" Fredric Remington

In 1540, Spanish explorer, Hernando de Soto, landed his expedition near the site of Tallahassee, Florida. While making his way to the Carolinas, in his never-ending search for gold and treasure, he became the first European to pass through the land now called the *Wiregrass Region*. This area runs along the modern borders of Georgia, Florida, Alabama, and Mississippi. It was a vast expanse of wild land, covered with 100' tall Longleaf Pines, Palmetto thickets, rivers, and more small streams and swamps than could be counted. De Soto, like many to come after him, saw the area as barren, with little or no value to King Charles or Spain, and continued his journey North.

As a Spanish explorer, he was also charged with colonizing the new world, and the King gave him four years to do it. When he first landed, he had six hundred and twenty men and 220 horses, to explore and colonize all of North America. Although the conquistadores failed to find much in the way of treasure before de Soto died in 1542, the Spanish were already busy staking out their first colonies in Florida.

The new settlers found life to be a struggle in the foreign terrain of Florida, but by the mid-to-late sixteenth century they were beginning to get a foothold in the alien environment and worked hard to expand Spain's empire. In 1565, Pedro Menendez de Aviles founded San Augustin (St. Augustine), and it became the first permanent settlement in the United States. He imported cattle to his newly established colonies to provide draft animals, hides, tallow, and a supplement for the colonists diets.

The Florida cattle were one of three basic groups of cattle introduced into North America by the early Spanish explorers. The first group of cattle (in no particular historical order), arrived at the California coastal regions as early as the sixteenth-century, but never moved very far to the east. The second consisted of Mexican (Spanish) feral cattle that moved north and in time mixed with wild local cattle to become what we know as Texas Longhorns. Like most cattle of the time, they roamed free, over millions of acres of prairie. When the locals first began to round up these wild cattle, they were slaughtered for their hides and tallow. After the population began to grow, they became an important food source.

The third group was from the original herds, introduced by Pedro Menendez de Aviles, and eventually migrated toward the trees and swamps of the *Wiregrass Region*. When de Soto first passed through the region, he failed in his search for gold and treasure. He didn't realize that he had found something that would eventually be considered a treasure — the abundance of thick, nutritious grass called wiregrass.

The wiregrass grew thick under the canopy of the longleaf pines and the well-watered open areas. Some cattle escaped the Spanish, and others were turned out intentionally, and many of them found the grass on their own instincts. For more than three-hundred years these cattle lived in the wild, reproducing and mixing with the occasional escaped cattle from other areas, eventually turning into a breed ideally suited for the area. Their small horns, lean bodies, lack of natural predators and the temperate climate of the South created a large, wild herd, available to those that could catch and hold them.

The cattle of this region were eventually called by many different names, including: Florida Cracker cattle, Criollo, Pineywoods, Romosinuano, Corriente and Florida Scrubs. In time, many of these were eventually bred for special purposes. The Corriente were developed into a breed that specialized in use for roping events in rodeos. Others, like the Pineywoods were used for their hides and tallow, as well as for draft animals, eventually being used as food.

The Spanish left the job of tending their vaca (cattle) herd to their vaqueros (cowboys), and they would be gathered as needed from the thickets and forests of the region. In time, as the settlers began to take up permanent residence in the region, they cleared land for crops and began to look at the wild cattle as a free resource if they could catch and hold them. By the time the British founded Savannah, Georgia, in 1733, there were tens of thousands of wild cattle living in the region.

Very few farmers had fences in those early days, if they did it was to keep the wildlife and the cattle out of their planted fields. The British brought with them a system of dealing with free-ranging animals. Men, referred to as pinders, gathered up any loose stock and locked them in their pinfolds (pens) and held them until the rightful owners claimed them by paying the fee charged by the pinder. Eventually, the wild cattle started to be a prime source of income, and were caught and held in the pinfolds for auction.

The local farmers found that they could be easily tamed and turned into draft animals. These were some of the first wild cattle, born in America, considered to be

Oxen. They were used for generations to plow, pull wagons, and even for skidding logs out of the forest when the timber industry was booming. In the beginning, they were used primarily as beasts of burden, and for their hides and tallow. As fencing became more common, farmers began to fatten them for use as beef cattle.

By the nineteenth-century, most of the original California cattle, imported by the Spanish were gone, or mixed with other breeds. Today there are efforts underway to resurrect the California breed from the remaining bloodlines. The Texas Longhorn is the largest remnant of the original Spanish cattle and is still popular today with breeders all over the country.

The original cattle of the *Wiregrass Region* are cousins of the Longhorn, sharing the speckled hide and general appearance with them. Due to the fact that they lived without any real human interference for more than three-hundred years, they evolved into a breed of tough, low maintenance, cattle that are now being bred to save those characteristics. Some prefer to call them pineywoods cattle, or cracker cattle or woods cattle, but all agree on one thing, these cattle found a perfect fit in the wild land of the wild *Wiregrass* and prospered on what Hernando de Soto and others thought to be of little value.

From Butchers to Barons

"Prospecting For Cattle Range" Fredric Remington

In September of 1850, a twenty-three-year-old German immigrant found himself on the streets of San Francisco in the middle of the gold rush. Legend has it that while in New York, he was offered a non-transferable steamer ticket to San Francisco with the name of Henry Miller on it. All Heinrich Alfred Kreiser had in his possession that day was his gold watch and six dollars in his pocket. Badly in need of a job, he took the ticket, became Henry Miller and found work in San Francisco as a butcher, the trade he was born into. In 1851 Miller opened his own butcher shop in the heart of the city.

His first venture into the live cattle industry was in 1854 after he found out that a herd of American shorthorn cattle had been driven from Salt Lake to the San Joaquin Valley. He purchased all three hundred head for $33,000 and moved them up to San Francisco. The average herd weight was about 800 pounds and the wholesale meat price ran about 18-20 cents a pound. Miller's first attempt produced a profit of more than $12,000.

Within a year Miller partnered up with another shop owner and German immigrant named Charles Lux. Their first partnership effort was to buy a herd of Texas cattle (considered a rarity in California at the time), slaughter and sell them for a profit. Although Lux had enough land for this venture, the drive for more of everything consumed both men. More cattle meant more land, more feed and more water. By 1858 the pair became permanent partners and their drive for more of everything was in full swing. Eventually the partnership, now called the *Pacific Live Stock Company,* owned land on both sides of the San Joaquin river for over a hundred miles.

From this moment forward, the livestock and meat industry of Northern California would never be the same. The two men were both aggressive entrepreneurs, with Lux taking the lead in the financial business and Miller understanding the cattle, the value of land ownership and the business of water.

Miller's passion was to be outdoors, looking for new land and managing the cattle business in person. He understood the intricate relationship of the cattle, the land and the water from the start. His drive for more was insatiable and his eye for details in the geography and drainages around the San Joaquin Valley taught him about irrigation. Eventually, the company would accumulate as much as 1.8 million acres in his lifetime. The Miller and Lux partnership was the single largest private landowner in the United

States in their day. Henry Miller became known as the *Cattle King of California*, a title that could have easily covered the whole U.S. at the time.

Miller built his home and offices near Gilroy, on the 1,700-acre Bloomfield Farm. He rarely went to the city and could oversee all his business from there. He preferred the company of his cattle and his cowboys to anything he ever saw in the city. He married Charles Lux's sister-in-law only to lose her in childbirth a year later. In 1860 he married her cousin Sarah, the marriage producing four children.

Lux was a city dweller from the start, enjoying the high society and San Francisco social life, only leaving the city long enough to conduct his business. He was the financial expert of the partnership, mixing easily with the upper crust of the bay area bankers and investors.

Miller and Lux began to accumulate land from the San Joaquin Valley to San Francisco. Guided by Miller's expertise, they bought private and public land and worked to consolidate their holdings into large tracts. Former Mexican ranches, lost in the American takeover of California of 1848, were sold by the current owners. Both buyers and sellers often took advantage of the previous Mexican owners and the flaws in the Land Claims Act of 1851.

The live cattle business continued to be the backbone of the operation. Each time more land was added, Miller made the improvements needed to run the next herd. If they were good in managing their money and land, Miller's innate understanding of the value of water was what made the whole operation a success. They bought up every riparian area possible and eventually secured the water rights along the San Joaquin, Kern, and San Benito Rivers. The partners invested in canals and irrigation systems, spending millions to bring water to the valley.

Their operation often butted heads with other ranchers and land speculators in the area. From the late 1870s until 1886 Miller & Lux were in court constantly over water rights lawsuits. Eventually winning the case on appeal, Miller knew the fight over water would never really end with all the new people wanting land and water. He made private deals with other companies to share the water and by 1888 (Charles Lux had died in 1887) had put together the largest irrigation network in the state of California.

Miller's operations were a natural for farming. The pools of unskilled laborers immigrants from Mexico and Germany, as well as white Americans, and a large Chinese work force were always available. The farms grew enough to feed all of its own stock and build up enormous stockpiles for bad times. He used local vaqueros, californios, and cowboys to manage his herds. One of his acknowledged skills was the ability to quickly evaluate and hire the best men.

The company maintained their own slaughterhouse and retail meat sales, butchering nearly a million cattle in 1912. They also raised thousands of sheep and had their own wool business. Miller had effectively created the first giant agribusiness in the country.

History records him as stern, straight to the point businessman. He was known be a tyrant when dealing with problem managers and was famous for his generosity to people in need. His workers were loyal to him and would do anything he asked without question. He started an annual feast called Los Banos Day that is still celebrated today.

Henry Miller died at his daughter's home on October 14, 1916. His estimated holdings at his death was 1.4 million acres of deeded land and control of another six million acres. His personal wealth was said to be about 40 million dollars.

Henry Miller and Charles Lux, two butchers that immigrated from Germany with nothing but ambition and desire for a better life, put together one of the most formidable businesses in American history. If Miller could come back for a moment, he'd be proud to see what part he played in the history of the area — and he'd be on the lookout for a few more cows and a few more acres.

Uncle Dick & Charlie

Raton Pass From a 1900 Postcard

For those people who like to say that there aren't any wild places left in the west anymore, it appears that they just don't know where to look. There are stretches of Southwestern America that are nearly as open and wild today, as they were when the earliest European explorers first saw them. One of these areas weaves its way east from Raton Pass, along the border of New Mexico and Colorado. From the pass lies a long stretch of foothills, canyons, and rivers that are still a wild land full of ancient mysteries and legends, a land once roamed by many of the giants of western history.

Raton Pass itself is filled with amazing history from its earliest days as a game trail and Indian path to its first modest expansion as the Mountain Branch of the Santa Fe Trail in 1822. This iconic western trail became a major delivery route for supplies to places like Fort Union and Santa Fe.

Uncle Dick Wooten

Charles Goodnight

In 1865, an ex-mountain man, trader, teamster and army scout, named Richard Lacey, aka *Uncle Dick Wooten*, began to see his chosen lifestyle disappear as settlers began to move into the country to the North. He decided to make trough and tumble frontier town, of Trinidad, just north of Raton Pass, his new home. Uncle Dick was well known long before his move to Trinidad. He counted Kit Carson among his close friends, and claimed to have killed the famous outlaw, Juan Espinosa, delivering a burlap sack containing the outlaws head to the Territorial Governor of Colorado for the $1,000 reward. Ever the entrepreneur, it didn't take long for Wooten to see a need to improve the pass and he knew a good opportunity when he saw it. After negotiations with New Mexico and Colorado, he formally leased the land from Lucian Maxwell, holder of one of the earliest land grants in the west and went into business.

The scale of the work was staggering, he had twenty-seven miles of road to improve, and many locals thought it was an impossible task. Uncle Dick hired a tribe of Ute Indians for his labor force and began work in earnest. "There were hillsides to cut down, rocks to blast and remove," said Wooten, "and bridges to be built by the score. But I built the road and made it a good one." Uncle Dick also built a good home on the pass, erected a toll gate, and settled into raising his family and operating the toll gate. Typical tolls were $1.25 for a single wagon or buggy, $.25 for a horse and rider and from $.05 to $.10 per head of livestock. The local Indian tribes had respect for men like Wooten and had many dealings with him over the years. He disliked the idea of killing Indians, and he showed his kindness toward them, when he pledged to make the road toll free to all the Indians who traveled it. Wooten kept his pledge and never charged them a toll for passage as long as he was in charge.

Not everyone was happy with the toll road, and some called it "Uncle Dick's highway robbery scheme." In time, most people came to understand that it was only fair to pay the man who invested so much time and money in the effort. The town of Trinidad, Colorado grew up with the early improvements to the trail and in later years claimed Bat Masterson as a city marshal for a time.

In 1866, after the pass was opened to toll traffic, one of those men unhappy with the new setup was a rancher named Charlie Goodnight, out of Texas. Goodnight and his partner Oliver Loving put together a large herd of cattle and moved them from Texas to Fort Sumner, New Mexico. In 1866, Goodnight moved another herd to Fort Sumner and decided to take part of it north to Colorado with an eye to the growing market in Denver. They drove the herd from Fort Sumner and up the Pecos River to Las Vegas (New Mexico). There they picked up the Santa Fe Trail north over Raton Pass and eventually to Denver.

In the fall of 1867, Charlie met Uncle Dick for the first time at the new toll gate. These two giant figures in western history must have had some spirited conversation, and after a long, loud protest, Charlie paid up, but he decided it was the last time he would pay the old mountain man to run his cattle over that pass.

By 1868 Charlie had made deliveries to Denver and Cheyenne and decided it was time to straighten out the trail north and avoid Uncle Dick's toll road altogether. He found a new route over Trinchera Pass, some 45 miles to the east. The new route was lower, better watered and easier to cross. The now famous route took him alongside Capulin Mountain, known to be an extinct volcano and designated a National Monument in 1916. The Trail, now commonly called the Goodnight Loving trail, also went by the tiny settlement of Folsom, later to become famous for archeological finds dating back more than 10,000 years.

Uncle Dick and Charlie lived through exciting and dangerous times. Outlaws and rustlers were common and disagreements with the Indians were still a real issue. The contributions made by these men and other pioneers of the day shaped the legends and the realities we know today. After a storied career of ups and downs in ranching, and a variety of other businesses, Charlie, age 91, married Corinne Goodnight (no relation) age 27, in March of 1927. Goodnight died in December of 1929.

As for Uncle Dick Wooten, his toll road ran until 1879, when the Atchison, Topeka and Santa Fe Railroad bought the right of way from him for a new rail line. They paid him and his wife a lifetime pension and he moved into Trinidad and lived there until his death in 1893. Along the way he had married four times, and had 20 children. By his death, he had outlived all of his wives and 17 of his children.

The foothills, canyons, and rivers are still a wild land of ancient mysteries and legends, a land once roamed by many of the giants of history.

Kansas City Here They Come

Kansas City Stockyards about 1900

Somewhere around the 1820s, a wandering French trapper named Francios Chouteau reached the confluence of the Kansas and Missouri Rivers. Setting up a small fort on the inside of a wide bend he named it Chouteau's Landing. In time, the spot would come to be known as the West Bottoms. It was a place long known to local Indians and even to Lewis and Clark, who had recorded that it would be a good place to build a small fort nearly twenty-years earlier. Chouteau prospered by trading with the Indians and stocked supplies for his business by riverboat.

By 1850, investors and entrepreneurs had built Westport Landing along the Missouri River. A group calling themselves the Kansas Town Company worked to get the settlement incorporated as Kansas, in Jackson County (Missouri). On February 22, 1853, it officially became Kansas City.

The location and timing were perfectly set for the little river town to become one of the most important locations in the history of America's western migration. The Oregon, California, and Santa Fe Trails all passed through Jackson County and the town provided the wagon trains with everything needed for the long trip west.

From the end of the Civil War until 1871, the town grew steadily from trade with the local Indians, agriculture, and livestock. In 1871, several local businessmen decided to set up a small stockyard in the bottoms. They brought in the first pair of Fairbanks scales in the area and built 15 unloading chutes and 11 pens. Built along the rail lines, records show that they moved more than 100,000 cattle from local producers to the rail cars in the first year. The new stockyards were built on the original site of the Central Overland and Pikes Peak Express Co. The company had gone out of business after the failure of its pony express operation and heavy competition from rivals.

The West Bottoms site sits squarely on the Kansas-Missouri state line, with a third of the site on the Missouri side and the rest in Kansas. The site was built to provide a centralized place where their cattle ranchers could sell their stock for the best price. Until the stockyards were built, they were forced to sell to the railroad at whatever price was offered. At the new exchange all cattle, sheep, and hogs as well as horses and mules were sold to the highest bidder.

By 1878, it had expanded to 55 acres from its original 13. Adding new loading docks and pens as well as sheds for sheep and hogs, it took up the whole West Bottoms and expanded across the river further into Kansas. Eventually, it grew to become one of the largest horse and mule markets in the country.

Kansas City was a natural transportation hub and as cattle herds began to show up from Colorado, Kansas, and other areas, new ranches flourished, and the flow of livestock was non-stop to the early railheads. The Kansas City yards continued to build more and more pens and understood early on the value of resting and feeding the cattle before they were loaded into rail cars and shipped to the packing plants farther east.

As the stockyard's operation grew, Philip D. Armour opened the first large slaughter house in the area. Records show they slaughtered 15,000 hogs and 13,000 cattle the first year they were open. By the end of the century, all of the big five meatpackers of the time had built plants in the area and worked hand in hand with the stockyards to become the largest meat suppliers in the region and the largest employers between San Francisco and St. Louis. By 1880, the city had grown to more than 60,000 people.

By World War 1, the stockyard's operation was selling hundreds of thousands of head of all types of animals every year. The meat packers were working at full speed to keep up with demand, helping to provide canned meat for the war effort. In September of 1918, nearing the end of WW1, the stockyards processed a record 55,000 cattle in one day and 475,000 for the month. In one three-month period over 1.3 million cattle came through the stockyards. If those numbers weren't impressive enough, as many as 34 railroads moved animals and meat in and out of the city. At one point, it was estimated that more than three million head of livestock were yarded in the city at one time. Livestock was the life's blood of Kansas City. Though many citizens hated the title of cowtown and the smell from the bottoms, the city thrived on it.

The yards had grown to more than 230 acres and employed 20,000 workers between the stockyards and the packer's businesses. The West Bottoms were receiving stock from 35 states and shipping to 42, processing about 170,000 animals a day. When the livestock exchange building was built, it claimed to be the largest and tallest building in the world.

Through the end of World War II, the stockyards continued to receive a million or more head of cattle a year for processing. Producers formed livestock cooperatives to group enough live cattle to fill whole rail cars for transport to Kansas City. Representatives of the different packers bought their animals at the exchange and refilled their pens daily.

After the end of the war, the picture began to change for Kansas City and the industry in general. New refrigerated trucks took advantage of the improved post-war roads and small sale barns became common. One of the biggest blows to sites like the Kansas City

Stockyards was the public's desire for grain-fed beef over grass-fed beef. It led to giant feed lots built close to the source of the feed and the packers also began to build close to the source of feed. As the packing houses in Kansas City began to lose money, they closed their doors. The big stockyard operations like Kansas City became little more than an interesting memory.

For 120 years the Kansas City Stockyards endured floods, fires, and World Wars and continued to sell and process tens of millions of animals for the public market and provide jobs for thousands of people. On September 27, 1991, progress finally caught up and it sold its last cow and closed its doors for good.

In 1990, a local investor named Bill Haw Sr. bought the old exchange building and remodeled it into an office building filled with tenants in different agriculture businesses, retail shops, art galleries, and restaurants. The only cattle left are in the pictures that line the walls.

A Town Named Wibaux

First Railroad into Wibaux, Montana

Up against the western border of North Dakota is Wibaux County, Montana (pronounced *"Wee-boh"* by the residents). Drive a few miles west from the border along highway 94 and you will come to a little cowboy town called Wibaux. Never more than a few hundred people, it services the agriculture industry in the surrounding area. The town had been through a lot in its early days, once named Keith, then Beaver and then Mingusville. In 1895, a cattleman and immigrant to the area thought Mingusville was not an appropriate name for his adopted home town and decided to change it in honor of himself. Starting a petition to change the name, he collected enough signatures to get it done despite some suspicions of impropriety by the locals.

Pierre Wibaux was born in Roubaix, France, in 1858, into a successful textile manufacturing business. He decided to leave France and the family business at the age of twenty-five for the great plains of America after hearing stories of great success in the cattle business. Reluctantly, his father, Achille, gave him a loan of $10,000 for a stake in his great adventure.

Traveling to Chicago, he engaged with people in the cattle business to learn all he could before investing his money. Meeting the infamous *Marquis de Mores*, a fellow Frenchman and adventurer shortly after he arrived, proved to be a stroke of luck for Pierre. The Marquis filled him with tales of the wild Montana prairie and the opportunities provided by the endless, grass-filled open range. Wibaux decided to stake his claim in the Montana Territory on Beaver Creek, building the W-Bar Ranch in 1883 and stocking it with Texas Longhorns. He moved his wife Nellie and son Cyril to the territory shortly after completion.

The stories the Marquis had told him had proven to be true, and within a few years he had accumulated a thriving herd of 10,000 head. The Marquis had a large open range operation in Medora, named after his wife, and Teddy Roosevelt owned the Elkhorn Ranch in the same area. For several years all the cattlemen of the area grew wealthy on the rich prairie grass.

Shortly after Wibaux, Roosevelt, and the Marquis started their operation the railroad reached the territory and the fortunes of the cattlemen continued to improve. The colorful Marquis left Medora for France in the fall of 1896 due to legal and personal problems and never returned. Wibaux and Roosevelt remained in the area and continued to do business as usual. However, the winter of 1886–1887 proved to be anything but business as usual for the cattlemen of the open range. The summer of '86 had been unusually hot and dry and a series of brutal snowstorms started early. By November huge snowfalls that had never been seen before combined with days and weeks of sub-zero temperatures.

It was not until the snow began to melt in the spring that people began to realize just how truly extraordinary the loss of cattle had been. Tens of thousands of bloated carcass lay scattered across the prairie and piled up in every gully and against every windbreak on the range. Many of the remaining cattle were in poor health and many cattle operations lost as much as 70% of their herd.

The spring of 1887 proved to be the beginning of the end for the traditional open-range cattle operations. The majority of the ranches were never able to recover causing huge numbers of bankruptcies. Pierre Wibaux was one of those ranchers, losing most of his herd to the same storms. The difference was that he saw a clear future for his cattle business that others didn't. Returning to France, he convinced his father and others to invest in his new cattle operation. He would stay in Montana and buy up all the remaining stock he could find in the area and create a new type of ranch, one that didn't depend on the open range. He would grow alfalfa and feed his cattle through the winter.

With fresh investor money in his pocket, he returned home and purchased all the cattle he could find for cheap prices from the other affected ranchers. Wibaux knew that the beef shortage would eventually raise the prices for the cattle he did have and the animals that had survived would be the toughest and most resilient ones of the herd. By the 1890s, he had amassed what was said to be one of the largest herds in the country estimated to be 65,000 to 75,000 strong as well as hundreds of horses.

As the open range business died out, homesteaders began to move west, setting up farms, ranches and businesses. Again, Pierre saw a need and became an investor in the area developing his land and the service industries of the town. He became the president of the State National Bank in Miles City and started his own bank in Forsythe. Wibaux was also owner and developer of the Clover Leaf Gold Mining Company which had found success in the goldfields of the Black Hills.

His success gave him the freedom to practice his philanthropy in both America and France. The French awarded him the *National Order of the Legion of Honor* for his help in building hospitals and model dairy farms to produce fresh milk for children. He invested in local businesses and built churches in the region, including *St. Peter's Catholic Church* in Wibaux. In the Black Hills gold mining region, they named the town that grew up around the mining camp Roubaix, after his home town in France.

Today, the little Montana cowboy town just across the western border of North Dakota is still taking care of the cattle business as it has done for generations. The high school's six-man football team is appropriately named the Longhorns and the pride in their town shows every time they take the field. The name Wibaux has served the town well for 123 years and counting.

Pierre Wibaux died in Chicago in 1913 and was buried in his namesake town. Standing on a giant granite pedestal in a small park is a twice life-sized bronze statue of a pioneer immigrant with a large hat, traditional buckskin shirt and chaps holding a Winchester in one hand and binoculars in the other stands Pierre Wibaux, proudly surveying all that helped build.

N. K. Boswell

"He Lay Where He Had Been Jerked" C.M. Russell

In 1853, Nathaniel Kimball (N.K.) Boswell, one of twelve children born in New Jersey to parents Lucinda and John, left home at age seventeen in search of his future. After several years working as a lumberjack in Michigan and Wisconsin, he met and fell in love with Martha Salsbury and they married in 1857.

 Several months later his crew was headed to an island to cut timber and the boat capsized in a sudden storm on the icy waters of Lake Michigan. Boswell's two coworkers died and he narrowly escaped with his life. Surviving a serious bout of pneumonia left him with damaged lungs and the doctors suggested a change of climate. He decided to head west, leaving his new wife until he found a place to settle.

In the Colorado Territory, he worked in various mines and the lumber business. Answering the call of the Union Army, in 1861 he signed up with the Colorado Volunteers. In June, 1864, he was a member of the regiment that responded to the horrific aftermath of the Hungate Massacre (near Elizabeth, Colorado), where Arapaho braves brutally murdered and mutilated a settler, his wife, and two infant daughters. A month later, his regiment was with John M. Chivington at the Sand Creek Massacre (near Kit Carson, Colorado, a retaliatory raid killing some 150 Cheyenne and Arapahos, mostly women and children and was wounded at the start of the engagement.

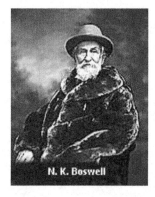

After the war, Boswell moved to Cheyenne, Dakota Territory. Finding the town to be little more than a small cluster of tents and shacks along the new railroad, he decided to stay and start a business based on the promise offered by the new railroad. Boswell sent for his ever patient wife Martha to join him. They became two of the earliest settlers and business owners in the town of Cheyenne (to become the Wyoming Territory the next year).

In the years following the Civil War, the territory became a haven for every kind of outlaw the West had to offer. The completion of the trans-continental railroad in May of 1869 saw the prosperity of railroad towns like Cheyenne and Laramie explode. Bank robbers, cattle rustlers, murderers and anyone looking to escape a bad past and make a quick buck flocked to the wide-open opportunities it offered.

With little or no law enforcement, the new railroad towns like Cheyenne and Laramie had become a favored hangout for many of the really bad men of the day. Soon Nathaniel's brother, George, and his family joined him and settled in to the wild little town.

Boswell held title to a claim for a Colorado gold mine and traded it to a failed pharmacist for his stock of drugs and opened the first pharmacy in Cheyenne. When he went looking for a druggist to run the operation, he had two applicants for the job. Hiring both of them, he started a second pharmacy in Laramie.

Boswell soon realized that the criminal element was seriously hurting the town's businesses. He joined a local "Citizens Committee" (vigilante) group to help combat the many gangs of criminals in the area.

Denver City Marshall David J. Cook had heard of Boswell's work in Cheyenne, and enlisted him in a group of crime fighters he called the Rocky Mountain Detective Agency. By the end of 1868 the detective group had shot or lynched most of the prominent desperados like Asa Cook, "Long Steve" Young, "Big Ned" Wilson and Con Wager. Cook shot Ed Franklin dead when he resisted arrest in Golden City. Sanford Duggan and gang leader Lee Musgrove were also captured and promptly lynched.

In 1869, Territorial Governor John Campbell appointed Boswell sheriff of newly formed Albany County that ran all the way from the Colorado border to the Montana Territory. Boswell also accepted the position as warden for the new Wyoming Territorial Penitentiary outside Laramie. One of the jobs of the sheriff was to collect taxes, a job he despised, as it meant days in the saddle collecting from every corner of the huge county. In 1872 he declined to run for re-election.

With a large following of admirers and friends encouraging him, they got him to accept the job as city marshal of Laramie. In August of 1876 he arrested Jack McCall, the infamous murderer of Wild Bill Hickok in Deadwood, and sent him back to be tried and hanged.

Convinced to run for county sheriff again in 1878, this time without the job of collecting taxes, he won easily. When the Black Hills gold rush started the criminal element raised its head again and new gangs were robbing the stages between Cheyenne and Deadwood. When word reached Boswell that two peace officers had been killed by outlaws, he raised a posse of 14 experienced men and took up the chase. They soon captured four of the key members of "Big Nose" George Parrott's gang without gunfire. Boswell did a little "creative" interrogation of the gang members until he got the location of Parrott and his partner "Dutch Charlie" in Montana and wired the local sheriff to arrest them. Before the trial, both men were dragged from the jail by a mob and issued their own version of "hempen justice" from the nearest telegraph pole.

Boswell was elected to his fourth term as sheriff in 1880, but by 1882 he'd had enough of the life and moved permanently to his ranch south of Laramie along the Colorado border. Working a large cattle operation kept him busy and out of the local politics. In July of 1883, the Wyoming Stock Growers Association started a detective bureau to combat the growing problem of rustlers. They offered him $200 a month if he would run it. The bureau's work against the outlaws was kept strictly confidential, but Boswell's creative methods of locating the rustlers worked well. It was reported in the 1884 annual meeting that Boswell's ". . . efforts had been remarkably successful." The trouble with rustlers had dropped significantly.

He retired permanently from his law enforcement work, and spent the rest of his life operating the N.K. Boswell Ranch with Martha and daughter Minnie. He continued to work to improve the ranch and his herd and generally stayed out of public eye. The transcontinental railroad had helped tame the once wild stretch of the frontier, and with the ability to ship his beef, he settled into the role of successful rancher. When Martha died in 1893, he was devastated and rarely left the ranch.

One exception was in 1903, when President Theodore Roosevelt came to Cheyenne and Boswell was one of ten honorary U.S. Marshals given the honor of escorting him on a horseback trip to Laramie.

Today, the N.K. Boswell Ranch, along the east side of the Medicine Bow Mountains and up against the Colorado border has been preserved as a representative example of a medium-sized ranch of the period. It was listed on the National Register of Historic Places in 1977.

Nat & Deadwood

"The Broncho Buster" C.M. Russell

Everyone loves a great *Wild West* story filled with six-guns, rustlers, Indians, horses and of course, great herds of cattle. There were a few old-time cowboys that wrote about their adventures in the old days, but most just did their job and quietly faded away into history.

Nat Love, (pronounced Nate) however, was never one to do anything quietly, and in 1907, published his autobiography, *Life and Adventures of Nat Love, Better Known in the Cattle Country as "Deadwood Dick," by Himself.* His tales of cattle drives, Indian battles, stampedes, storms, and personal heroics could easily make a dozen Hollywood westerns.

Born in 1854, as a slave on a plantation in Davidson County, Tennessee, he was raised in a typical slave cabin with two siblings. Nat's father, Sampson, was a slave forman and his mother worked in the "big house." In the antebellum south it was strictly against the law for slaves to learn to read and write. Although neither Nate nor his father had any formal education, his father disregarded the law, risking severe punishment, and helped him learn in secret.

In February of 1869, he left Tennessee with fifty dollars received from breaking colts and the sale of a horse, and headed west for Kansas Territory. "I wanted to see more of the world, and as I began to realize there was so much more of the world than what I had seen, the desire to go grew on me from day to day."

Love walked, hitched rides, and eventually worked his way to southwest Kansas. "It was in the west, and it was the great west that I wanted to see." Looking for work on a ranch near the new town of Dodge City, Love met up with a Texas cattle company that just delivered a herd and was preparing to head back. Never shy, Nat approached the trail boss and asked for a job. He had no proper trail gear, or no guns, and looked more like a farmer than a cowpuncher.

Nat Love

Love writes that the boss challenged him to ride a wild horse named "Good Eye." If he could ride him, he'd hire him. If he couldn't, there would be no job. Nat's account says he rode the horse until it tired out. When he dismounted, the boss hired him at thirty-dollars a month and took him to town and bought him complete new trail gear, including ". . . a fine .45 Colt revolver."

Love notes that his new boss gave him the nickname, "Red River Dick," which he carried for a long time. "The outfit of which I was now a member of was called the "Duval Outfit," and their brand was known as the Pig Pen brand."

On the trip back, he writes that they were suddenly attacked by Indians:

". . . there were about a hundred painted bucks all well mounted. They were coming after us yelling like demons." Their fifteen-man outfit lost one man in the fight, most of their packing gear and all but six horses. *". . . Although we had the satisfaction of knowing we had made several good Indians out of bad ones."*

Nat worked as a trail hand in some of the earliest recorded drives, through rough country, rougher weather, cattle and buffalo stampedes and Indian attacks. He drove herds to places like Deadwood, in the Dakota Territory. His book tells of winning $200 in the 1876, Deadwood 4th of July rodeo. It was here that he says was tagged with the nickname "Deadwood Dick" after winning a shooting contest. Throughout that period the name was used by many others, including in popular dime novels of the day.

For all the braggadocio in the book, he was a genuine working cowboy for more than twenty years. The Duval Ranch was on the Palo Duro River and Nat stayed on for three years delivering cattle and horses as far as the Dakotas and Wyoming.

From Texas, he took a job at the Gallinger Ranch in southern Arizona, working for them from 1872 through 1890. Love drove large herds of cattle and horses to markets all over the West. "I soon became one of their most trusted men, taking an important part in all the big round-ups and cuttings throughout western Texas, Arizona and other states where the company had interests to be looked after."

Love's account of the Wild West is one that fans of the era would find fascinating. It covers the time from the early cattle drives, and when the buffalo numbered in the millions, to the arrival of the railroads. In twenty years, he saw his cowboy profession disappear, as well the bands of wild, ranging Indians he battled and the great herds of buffalo. "Buffalo hunting, a sport for kings, thy time has passed . . . where once could only be heard the sharp crack of a rifle or the long doleful yelp of a coyote," lamented Love.

By 1890 the writing was on the wall, the adventures on the open prairies was over and he was ready to move on. "To us wild cowboys of the range, used to the wild and unrestricted life of the boundless plains, the new order of things did not appeal, and many of us became disgusted and quit the wild life for the pursuits of our more civilized brother. I was among that number and in 1890 I bid farewell to the life I had followed for over twenty years."

A year later Nat Love travelled to Denver and hired on to the Pullman service on the Denver & Rio Grande Railroad. After marrying a woman named Alice, they moved into a cottage in Denver and Nat began his job as a porter on the route between Salida, Colorado and Denver.

Love talks about meeting many of the iconic characters of the period including Bat Masterson, Buffalo Bill, Yellowstone Kelly, Billy the Kid, the James brothers, Kiowa Bill, Pat Garrett and many others. His story is remarkable, even though he had been accused of stretching the truth on a few occasions. He writes of being shot fourteen times over his years on the trail and being captured by Indians and then escaping at night on a stolen Indian pony.

Love wrote with a view point that was popular in the period, claiming men like Frank and Jessie James gave their stolen money to needy people and that Billy the Kid was an honorable man.

Even with some bragging and rationalization of the times, Nat and Deadwood left an incredible record for the world to see it as it was through the eyes of someone who was there.

Nat and Alice had one daughter, and retired to California. He died there in 1921.

Charles Littlepage Ballard

Teddy Roosevelt Front Center and Charles Littlepage Ballard behind him

Charlie Ballard was born October 11, 1866, just eighteen months after the Civil War, on a ranch in Hayes County TX. The first child of ex-Confederate soldier Andrew Jackson Ballard, and his wife Cathryn Elizabeth, Charlie would eventually have seven more siblings.

In 1875, Charlie's dad was looking for new opportunities to make money and moved his family to Fort Griffon, in Shackelford County. Located on the rough edge of civilization, he planned on becoming a buffalo hunter. The fort and the nearby town catered to hunters, and was one of the wildest places on the frontier. For the next three years he made his living in the buffalo trade.

Not long after Charlie's ninth birthday, his father sent him out to find a missing wagon team that had pulled their pickets and ran off. Running into a group of Indians that were dancing, singing, and celebrating loudly sent him racing back to the fort to tell his story. It turned out that the first "Wild Indians" he'd ever seen had recently procured a supply of whisky and were just happily drunk. Charlie loved recalling the story of how the wild Indians scared him that day.

In 1878, the Ballard clan was headed west again. "The buffalo had played out and became so scarce that it was unprofessional to kill them for their hides," said Ballard years later. "We left Fort Griffon in February . . . There were no further settlements until we got to Fort Sumner, New Mexico, (nearly 400 miles). The only thing on the plains at that time was a few buffalo, wild horses, antelope, and a lot of renegade Comanche Indians from the Fort Sill Reservation in Oklahoma."

Charles L. Ballard

The Ballards made the trip in three weeks. They found the fort abandoned when they got there. "The houses were still in livable condition and there were thirty or forty families living there. We moved into a house back of the old suttlers store."

Still looking for prosperity for his big family, Ballard moved his family to Lincoln and Charlie got to know all of the principal characters of the Lincoln County Wars including, Billy the Kid, the Dolan Gang and well-known one-armed attorney, Houston I. Chapman, murdered in cold blood on the main street of Lincoln.

Soon the family was on the move again, settling in Roswell fifty miles to the east. On a night trip home from Lincoln to Roswell, Andrew Jackson Ballard was shot in the chest while in his buckboard. The bandit also killed his horse and Ballard was able to walk five miles for help. The bullet eventually cost him his life and the shooter was never found. Charlie, now fourteen, had to step up and take the reins of the big family.

Charlie went to work for Captain J.C. Lea as a cowboy and teamster on his ranch. *". . . the best friend I ever had and one of the finest men I ever knew."* Said Lea, about his friend.

In 1888, he married Araminta (Mintie) Corn and they would eventually have six children. Charlie was a small stock raiser and was elected city marshal of Roswell, and appointed deputy county sheriff at the same time. In 1895 he was commissioned U.S. Marshal for the Territory of New Mexico. By 1896, the New Mexico and Arizona territories became a haven for outlaws and Charlie was continuously involved in tracking and capturing many bad men of the day.

On February 15, 1898, the U.S. fighting ship, the Maine blew up in the Havana, Cuba harbor and two months later the U.S. declared war on Spain. "I volunteered and helped raise recruits, taking twenty-five men from Roswell," wrote Ballard in his memoir. He was mustered in at Santa Fe as second lieutenant and shipped immediately to San Antonio. There they joined up with Colonel Leonard Wood and Lieutenant Colonel Theodore Roosevelt.

On Ballard's muster out papers, Colonel Roosevelt wrote, "This officer I regard as one of the very best in the regiment; I know none in whom I grew to place more trust, and none whom I would be more anxious to have with me if called again into the field."

After he mustered out, he returned to the ranch in Roswell. By 1899 the government wanted him again, this time to fight the rebels in the Philippines. He accepted a commission and sailed from San Francisco. Contracting health problems, he went back to the ranch in early 1900.

In 1907, the territorial governor appointed him to the Territorial Cattle Sanitary Board. As his business flourished, he acquired more land and hired Pat Garrett to manage one of his operations, the Salt Springs Ranch.

In 1903, Ballard's natural curiosity and love of travel caused him to accept an invitation from his old friend, D.M. Goodrich (Goodrich Rubber), asking if he would come along on a trip to South America. *". . . I cabled him that I would, not knowing where I was to go, or why."*

Ballard met with Goodrich and listened to a story told by railroad builder Archer Harmon, who was building the transcontinental railroad across the Andes Mountains. He told them of the Galapagos Islands and herds of wild cattle covering the largest island. The men wanted him to scout it out for a possible cattle ranch.

Agreeing to the proposition, Ballard and one fellow cattleman landed on the island in early 1903. Finding a few saddle mules to rent, they spent thirty days exploring. "The island was seventy-two miles long and forty miles wide. I figured there were about thirty thousand head of cattle." After being abandoned on the island for another month, they ended up eating mostly tortoise and calf meat. When Ballard returned to the Ecuador mainland, he began to search for who he could get title and ownership from for the cattle and land. Before the deal was done Ecuador had a takeover revolution and the dream of an island ranch ended.

In New Mexico, his ranching business was prospering and he was elected to the legislature in 1905. The same year he returned to Washington D.C. as a guard of honor for President Theodore Roosevelt. Ballard went on to explore Guadalupe Island, a Mexican territory for the possibility of a cattle ranch. Although he found some good timber and thousands of fat goats, he decided against it as a cattle island.

In 1912, New Mexico gained statehood and Ballard found more success in farming and land development. Eventually, a long, disastrous drought killed off thousands of cattle and destroyed crops across the state. By 1924, Ballard was wiped out financially. Charlie refused to take bankruptcy and sold everything he owned to pay off his creditors.

Charles Littlepage Ballard experienced the West and followed his love of adventure as few men of his time did. Ballard passed on April 17, 1950 and is buried next to his first wife "Mintie" in Evergreen Cemetery in El Paso, Texas.

"We Pointed Them North"

"Trailing Texas Longhorns" C.M. Russell

In the summer of 1937, Helena Huntington Smith, a writer and lover of the old West, was searching for some authentic background material for her novel about the cowboys and cattle drives of the 1870s and 1880s. Fortunately for generations of Western history enthusiasts, her search led to an old-time cowboy and cattle rancher by the name of **Edward Charles "Teddy Blue" Abbott**.

She tracked him down and found him living with his wife, Mary, on a ranch outside of Lewistown, Montana. Smith wrote that "Today he is seventy-eight years old, tough as whipcord, diamond clear as to memory, and boiling with energy." The timing was fortunate for both of them. Abbot had been trying to write a book of his own adventures without much success. After spending time with him, she set aside her own novel and offered to help write his story. The result: *We Pointed Them North: Recollections of a Cowpuncher*, published in 1939, is considered one of the most respected histories ever written of the cattle drive period.

Born in Norfolk County, England, in 1860, Teddy's father packed up the family and moved to America in 1871. Abbott's father, J.B., took up ranching and farming just outside Lincoln, Nebraska. Soon after he settled the family in, J.B. made the long trip to Texas to put together a herd of longhorns.

"I was the poorest, sickliest little kid you ever saw," wrote Abbott. *"All eyes, no flesh on me whatsoever."* He was taken along with the hopes that the fresh air and time spent in the saddle would toughen him up. Abbott told Smith that the days of hard work on that first drive, *"made a cowboy out of me. Nothing could have changed me after that."*

Teddy Blue Abbott

That first herd of longhorns got J.B. started in the cattle business and provided the perfect playground for a young man to learn the business and live out his dreams. Teddy's dream had started out not to be a cowboy, but rather an Indian. They were in Pawnee country and he spent all the time he could with them, but try as he might, he could not get them to accept him as a member of tribe. After being sent away for the last time, he changed his mind and decided a cowboy career was the next best thing. Either way, he could still live the nomadic life he dreamed of. Instead of following the buffalo herds, he would be moving cattle along the same trails.

"There were worlds of cattle in Texas after the Civil War," wrote Abbott. *"By the time the war was over they was down to four dollars a head — when you could find a buyer. Here was all these cheap, long-horned steers overrunning Texas; here was the rest of the country crying for beef — and no railroads to get them out. So they trailed them out, across hundreds of miles of wild country thick with Indians."*

Helena Smith worked with Abbott through '37 and '38 and was enthralled at his tales of the cowboys and the cattle drives. Teddy's motivation to get the story down was clear. He wanted to set the record straight, to tell the story exactly as it was. After reading the accounts written by others, he told her they may be accurate, but were basically plain and boring. "They never put in any of the fun, and the fun was at least half of it."

He hated the politeness and the cleaned-up versions of their text. "You have all these old fellows telling stories, and you'd think they were a bunch of preachers, the way they talk. And yet some of them raised more hell than I did."

Abbott talked with his friend Charlie Russell about making a book together, with Charlie doing the pictures, but Russell died in 1926, and the book was put away and forgotten. Helena Smith realized that Abbott would not be well served if she tried to clean up his prose for polite society, so she wrote it exactly like he told it, all realism and little romanticism, with all the salty language and trail dust included. Abbott recalls:

"Those first trail outfits in the seventies were sure tough. They had very little grub and they run out of it and ate straight beef; they had only three or four horses to the man, mostly with sore backs, because the old-time saddle eat both ways, the horses back and the cowboy's pistol pocket."

By age 19, his father and brothers got out of the cattle business for other ventures, and Teddy found himself in Austin, Texas, signing on with a questionable company known as the Olive outfit. They drove a herd 800 miles to the Loup River country in Nebraska. Averaging 10 to 15 miles a day, the drive had every bit of adventure that a young cowboy could ever want and then some. The drive had stampedes, Indians, rustlers, floods and long stretches without water. It made for long brutal days and great stories.

Teddy's accounts of drinking, bar women, practical jokes, and cravings for things they missed on the trail (like oysters and celery), are classic. His eye for the detail and his sharp memory described the men of his time, *"In person, the cowboys were mostly medium sized men, as a heavy man was hard on horses."* He explained that no self-respecting cowboy would ever carry two sixguns. His descriptions of the clothes, customs and vernacular of the day is amazing. The detailed accounts of Dodge City, Kansas, Ogallala, Nebraska, Miles City, Montana and other cattle towns of the day give us a unique look into life before modern civilization closed in.

Teddy Blue was a product of the period he lived in. He was an outspoken atheist, claiming "Ninety percent of them (cowboys) was infidels." He hated the "nesters" coming west to put up fences and carve up the land. He loved the bare-naked prairie for what it was, a beautiful, empty paradise with nothing but buffalo, Indians, and grass.

When backstage at a theatre in Miles City, he caught his spur on a carpet and crashed through a partition and onto the stage. Playing it for a laugh, he straddled a chair and bucked it across the stage hollering "Whoa, Blue, whoa, Blue . . ." Teddy Blue was born at that moment and stayed with him for the rest of his life.

After several more trips north with Texas herds, he married Mary Stuart in 1889, the half-Shoshone daughter of legendary Montana pioneer, Granville Stuart, and settled in as a cattleman near Giltedge, in Fergus County, Montana. The couple raised five boys and three girls and became active in local politics and eventually had 14 grandchildren and one great-granddaughter.

Teddy Blue died of pneumonia in 1939, just days after the book written by Abbott and Smith was published.

The Fever

"Worked Over" C.M. Russell

Chester A. Arthur, became America's 21st President after the death of James A. Garfield from an assassin's bullet. Arthur, who was a Civil War veteran, a lawyer and lifelong politician, had served in many different political patronage jobs in New York before being elected Vice President on March 4, 1881.

After a long career of working in nothing but well-paid political appointments, he surprised everyone by declaring himself a reformer. Most of his old political cronies laughed out loud at the idea of him changing how business was done in Congress. One of his first major tasks in office was to take on the civil service. By signing the Pendleton Act, a bill that required jobs to be assigned by merit and not patronage, he proved he was serious about his new-found reformer status.

On May 29, 1884, he signed a bill creating the Bureau of Animal Industry (BAI), organized under the United States Department of Agriculture. They were charged with preventing diseased animals from entering the food chain. The new agency replaced the Veterinary Division that earlier had replaced the Treasury Cattle Commission.

The country was rapidly expanding, and from the start the bureau was nearly overwhelmed from having to deal with many contagious diseases like cattle lung plague (bovine pleuropneumonia or CBPP) and other health issues both contagious and non-contagious. The earlier agencies were generally poorly funded and lightly staffed. In June of 1868, *Veterinarian,* an English agriculture journal published one of the first accounts of what would become the countries next great cattle disaster.

The report stated that in Illinois, a new danger to the industry had been discovered. Cattle producers of the area stated that: *". . . a very subtle and terribly fatal disease"* had broken out among cattle herds in Illinois. The disease killed quickly and was reported to be, *". . . fatal in every instance."*

By the time the BAI was formed, the Texas cattle industry had been trailing large herds of open range longhorn cattle to all points on the map for nearly twenty years. A whole new problem was ready and waiting when Doctor Daniel E. Salmon took the helm of the new agency. He had studied at Cornell University and earned the first Doctor of Veterinary Medicine degree ever awarded in the United States. It proved to be one of the most important positions ever filled under the Pendleton Act. By the time he became director, he had been working on diseases like lung plague for years and was famous in his field. One of the diseases he had co-discovered and studied extensively, was salmonella, named for him in honor of his work in the field of infectious disease.

Salmon knew that one of the most serious issues in the American livestock industry for years was the importation of animals from foreign countries. Diseases like lung plague had been introduced many years before. It had been more prevalent along the eastern seaboard, but the cattle operations in the east were run on much smaller properties and were easier to control.

The West and the Midwest proved to be a challenge for Salmon and his team on a different scale. So much of the land was wide open and unfenced, that hundreds of thousands of these range bred longhorns were being trailed all over the country completely unrestricted. For years the states north and east of Texas had been trying to stop the advance of a disease that had been killing the local cattle whenever the longhorns moved through their areas.

A new battleground had opened up in the West, and it was all about a tiny little eight-legged terror called a cattle tick, or Lone Star tick, one of nearly a hundred types in the continental U.S. At a quarter inch long, this ugly, brownish-green little creature had a love affair with the blood of the longhorn. Geography and climate had kept it generally in the Southwest and the South, the natural range of the wild longhorns.

Early on, the panhandle of Texas was one of the first areas to vocally resist the trail drives from the south. As the cattle moved through new parts of the country that historically didn't have the problem, the ticks, now fully engorged with blood would fall off and rapidly propagate. As the drives went farther north and east to railhead cities, the local cattlemen knew they had to stop the longhorns from entering their country. When they heard the drives were getting close, they closed ranks and put up what they called the Winchester Quarantine. More than one battle was fought between the cowboys over the presence of the southern cattle. Out of frustration and anger, everyone called it *Texas Fever* (Babesia).

Salmon and his team knew that the single host for these creatures was the longhorn. They began to work on a formal quarantine line, where the cattle couldn't be moved north during the prime tick breeding period. The BAI instigated many new measures to start to eradicate the problem.

In some cases, the cattle had to be slaughtered because the disease was too far advanced. The bureau advocated burning the range, pasture rotation, spraying the animals with a chemical to kill the tick, treating the range, and moving them through a chemical dipping process. The procedures initially started by Dr. Salmon and the BUI are still being continued and improved to this day.

The BAI improved the existing quarantine stations on all ports of entry and created new ones. As the country expanded, the duties of the bureau increased dramatically. Divisions were created for animal husbandry, animal nutrition, animal pathology, dairy and zoological studies. They set up research divisions for disease and parasite study, as well as animal and meat inspection.

In 1905, Upton Sinclair published his novel/expose called *The Jungle*. The book exposed the horrific conditions of the meat processing business and the equally terrible treatment of the workers in Chicago's meat packing industry. The outrage caused President Roosevelt to call for an investigation, resulting in Congress creating the Federal Meat Inspection Act of 1906.

The BAI now had the responsibility for the safety of all the meat consumed in the U.S. They immediately created stiff new sanitation requirements for the meat packing businesses, and mandatory inspection of the meat before killing and before the carcass leaves the plant. They imposed strict sanitary standards for all slaughter houses. They hired hundreds of inspectors to work at 163 kill plants around the country.

The Bureau of Animal Industries ran independently from 1884 to 1942 when it became part of Agricultural Research Administration. In 1953, President Eisenhower reorganized the USDA and the bureau became part of the larger program. Doctor Daniel Elmer Salmon, hired by a new president, under new laws, proved to be the right man at the right time to safeguard the health of our growing population.

Utah's Last Cattle King

The Gather © Bert Entwistle

Preston Nutter, born in what is now West Virginia in 1850 was the son of a well-known horse breeder. He lost his father when he was 9 and his mother shortly after. Sent to live with relatives he disliked, he ran away from them at age 11. Working his way west he found himself at the Mississippi River and found a job as a cabin boy on a riverboat. Still restless, he joined up with a wagon train in 1863 and headed west. Nutter would forever mark the trip by choosing "63" as the brand for all the horses and mules he would ever own.

As the group prospected east into Colorado, he soon realized that Packer was a phony, knowing little about prospecting and even less about the Colorado mountains.

Calling him a "whining fraud," he and several others left the group and spent the winter with Chief Ouray and the Utes near Montrose, Colorado. Packer, and five other members of the group, pushed on into the now snow-covered Rocky Mountains.

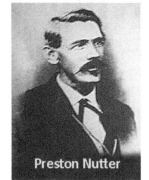
Preston Nutter

In the spring, he was back on the hunt for his next adventure and settled into Colorado, serving a term in the Colorado legislature. With politics out of his system he went into the freighting business, hauling ore from local mines to the mills — a much better deal than digging it himself.

In 1874, Nutter found himself in the Lake City, Colorado courthouse as a witness against Alferd Packer for the murders of the five men he took into the mountains when the party split in Montrose. Worse, he was said to have eaten the remaining members of the party. Packer was found guilty and legend has it that Judge M. B. Gerry pronounced his sentence with these famous words:

". . .Stand up yah voracious man-eatin' sonofabitch and receive yir sintince. When yah came to Hinsdale County, there was siven Dimmycrats. But you, yah et five of 'em, goddam yah. I sintince yah t' be hanged by the neck ontil yer dead, dead, dead . . ."

Nutter was to make the trip back to Colorado at least once more, after Packer escaped and was re-tried a second time. By 1886 Nutter had expanded into the cattle business buying a herd of cattle and setting out to find suitable range to run them on. He decided to head back to Utah, where he had seen good grazing ground on the gold mining trip. He drove his herd west into their new home between Thompson Springs and Moab. Not long after he established himself, he made a deal with the Cleveland Cattle Company to trade a 1,000 head of his mixed breed cattle for the same of Cleveland's purebred Herefords. Many western ranchers didn't like the English breed, and the Cleveland outfit was happy to get rid of them. Nutter, with remarkable foresight for the time, had already decided that the Hereford would become the new breed of the West and staked his future on it.

He found enough ground to accommodate as many as 20,000 head in the Thompson's Spring area in the northeastern corner of Utah. Nutter did something nearly unheard of in the days of open range ranching — he introduced supplementary feeding to his herd. Before winter set in, he moved his herd closer to the winter range around his ranch and nearby rail spur so he would always have access to a source of feed and shipping . This new idea saved him from the horrific cattle loss that most open range outfits experienced in the great blizzard of 1887.

A year later he formed the Grand Cattle Company with two partners. For the next few years, the ranch flourished and the herd grew rapidly. Grand Cattle bought out many smaller operations that had been devastated by years of droughts and blizzards. The expansion brought him some of the best grazing land and good springs in the area.

The 1893 economic collapse, often called the Silver Panic, caused great losses in the industry, but it wasn't as bad for the conservative Nutter. When most outfits were hurting, he was able to negotiate a deal with the government to lease the Strawberry Valley, along the west edge of the Uinta Basin. This gave him another 665,000 acres of prime range, and he began to advertise as far away as New Mexico and Arizona for cattle to stock it.

Eventually, he set up an operation near St George, for the warmer climate. Throughout his ranching career, he fought the weather, politicians, bandits, and Indians. He once wrote: *"I am plagued by rustlers, bootleggers and sheepmen."* He had many notable encounters with the sheepmen and tried to settle many of them in court, but it was a never-ending battle. He was among the first to openly support the Taylor Grazing Act, to promote the orderly use of public lands.

Nutter was famously a hands-on owner, spending days and weeks on horseback, policing his operation. He had been too busy for a family for the first 58 years of his life, but finally got captured by a woman 21 years his junior, named Katherine Fenton.

Fenton was a telegrapher, and moved west taking a job as the manager of the Colorado Springs postal telegraph. In 1905, the Uinta Basin was formally opened to homesteaders and she entered a lottery for a chance at a claim of her own. She won a homestead in Ioka, Duchesne County, Utah Territory. Taking the stagecoach to see her property, fate took another turn of good luck when the stage driver got lost and they spent the night at Preston Nutter's ranch. She continued to work as a telegrapher for several more years and shared her time between her Utah property and the Nutter ranch, finally marrying Preston in 1908. They had two daughters, Catherine, and Virginia. It was often said that Nutter had a herd so large that there was no way he could possibly know how many cattle he had. Later in life, his daughter Virginia wrote:

"... He was much too shrewd a businessman not to have a reasonably accurate tally, but cattlemen are, by tradition, a silent lot when it comes to discussing their business."

The history of the Nutter operation is filled with stories of Indian conflicts, rustlers like Butch Cassidy and "Gunplay" Maxwell, as well as never ending battles with the sheepmen in and out of court. By the end of WWI, the decline of the large western ranchers began a serious decline. His operation lasted through the Great Depression when beef prices became so low that the old way of life was over.

Preston Nutter died on January 28, 1936. One of the Salt Lake papers announced it like this:

"One of the last links between the old West and the new. Wednesday afternoon will be buried Utah's last great cattle king, Preston Nutter."

The Unassigned Land

The Start of the Oklahoma Land Rush

Something called *Indian Territory* must have been a difficult concept to understand for the Cherokee Indians of the Southeast United States in 1838. The land where they live had been settled by their ancestors many generations before. Why would they want to live in this other place they had never been to when they already had everything they needed in Georgia.

To the United States, the land to support new immigration was more important than anything the Cherokees might want. They forced the Indian nations in the area to cede their land to government, (Treaty of New Echota / Dec.1835) in trade for new land in the Northeast corner of the Oklahoma Territory referred to as Indian Territory. The right to this land was granted in perpetuity by the new treaty. The maps of the time called it unassigned land. What followed was one of the most shameful events ever committed by the United States, the forced relocation of the entire Cherokee nation, it became known as *The Trail of Tears*. Other relocations had been going on since the early1830s, but this is considered by historians to be one of the worse acts ever committed against Native Americans. More than 16,000 Cherokee men, women, and children were relocated. Disease, starvation, and exposure on the trail claimed more than 4,000 lives. The Creek, Choctaw, Chickasaw, and Seminole (with the Cherokees, they were often called the five civilized tribes) had been relocated several years before.

The government took the land they wanted and considered the relocation of the Indians to be an experiment in trying to civilize the tribes. They hoped that a successful ranching business in the Oklahoma Territory would lead to things like Indian-owned railroads and other businesses and to eventual self-sufficiency.

The Indians were already accomplished cattle ranchers and farmers, but they never needed or wanted the other businesses. The five tribes brought hundreds of cattle with them on the move west, and the cattle industry in the territory began to prosper. The five nations combined to share the vast, open range grasslands assigned to them.

By the 1850s, news of the gold strike in California opened the door for thousands of gold seekers and immigrants to pass through the Indian Territory. Until then, the tribes had little in the way of an outside market for their cattle. As the numbers of immigrants increased, the need for the Indian cattle grew dramatically. For twenty years the five tribes built and maintained one of the largest cattle operations in the county. At the start of the Civil War in 1861, the Cherokees alone owned about 250,000 head.

Part of the land given to the Cherokees was a piece called the Cherokee Outlet, often mistakenly called the Cherokee Strip which is in Kansas. It was an enormous piece of the unassigned land about 225 miles long and 60 miles wide, west from the tribal resettlement lands, and north along the border of Kansas. The treaty gave this spectacular piece of natural grazing land to them the for their grazing and buffalo hunting.

By the end of the Civil War in 1865, the big Indian nations had been devastated. The five nations together had lost over 300,000 head of cattle due to theft and the chaos of war. The Cherokee Outlet, still Indian Territory at the end of the war, began to be used as a trail drive route for tens of thousands of cattle from Texas without any compensation to the five nations.

At the same time the government relocated several more tribes to the eastern end of the outlet. After the war, the increase in cattle theft became rampant and the outlet became known as a wild, no-man's land, serving as a base for many of the worst criminal operations of the day.

The economy of the five original tribes had been devastated. Indian Territory had been under the flag of the Confederacy until the surrender, and also lost thousands of their cattle to the southerners. As the United States began to go through reconstruction, the Texas range cattle industry began to flourish. A few individual tribe members fought to start their own ranching operations in later years. Wilson Jones, a Missouri Choctaw, started a ranching operation in 1867. Within a few years he had a successful operation of more than a thousand head. In 1890, he was elected Chief of the Choctaw nation and served three terms, and ran a successful ranching operation until his death.

At the end of the war, thousands of men, many of them former soldiers, had taken up the cowboy profession, working jobs in the new cattle industry of Texas. Wild longhorn cattle were plentiful, and ranchers had begun to build up large herds, but places to sell them in quantity was badly needed. Several early trail drives were attempted to places like Sedalia, Missouri, with marginal results, but in the Spring of 1867 the railroad had reached Abilene, Kansas, and the cattle business and the outlet was changed forever.

As the trail drives of the Texas cattle began in earnest, the outlet, still owned by the five nations, sat squarely between the cattle owners and the cattle buyers. It was either go through Indian Territory or go hundreds of miles out of your way to go around it, the decision was an easy one for the cattleman.

The outlet was well known for the quality and quantity of its grass and water. By the time Abilene was shipping cattle by train, the Texas herds were crossing it in record numbers. Established ranchers from both Oklahoma Territory and Kansas routinely grazed their herds on the outlet to the point that tens of thousands of cattle could often be seen feeding at the same time.

The Cherokee still had rights to the land by their original treaty, but the future was easy to see. The flood gates had opened and nothing could hold the cattlemen and immigrants back. The tribes filed grievances with the government and tried to police it themselves, even trying to collect fees from the trespassers, but it was an impossible task.

By the 1890s, the outlet had been completely closed by the government and the tribes were not compensated and not allowed to use the land. Having little choice, they agreed to sell more than six million acres from $1.40 to $2.50 an acre. On September 16, 1893, the Cherokee Outlet Land Run took place, and thousands of people claimed their own place on the outlet, ending forever the wild, open range that had supported thousands of Indian cattle.

About Bert Entwistle

I've been a history lover and a voracious reader as long as I can remember. As a boy, I read everything I could get my hands on. If it was about history, particularly American history, I was a happy kid. If it was about the Wild West, it was even better and I'd read it over and over again. Like most people in our little town in the 50s, we had a small, black and white television and two channels. We could get a few westerns, and Gunsmoke was my dad's favorite, so it was mine too. The more books I read, the more interested I became in the expansion of the country and the people that settled it. Although I loved Hollywood's version of the wild West, the real stories of those early pioneers I read about were much more interesting than anything I'd ever seen on television or in the movies.

After our two sons were grown and gone, I started working as a freelance photographer and journalist. After twenty-five years working for magazines and other outlets, and publishing hundreds of feature stories and columns in dozens of publications, I decided it was time to slow down. I now do only an occasional feature story or photo job, but continue to do a regular history column for *Working Ranch Magazine* called *Looking Back*.

The 50 columns included in this book came from my research and reading many first-person accounts written by the principals themselves. My man cave is buried in old books, pictures, and trinkets of the Old West — and I wouldn't have it any other way.

I love the art of the West as painted by Charley Russell, Fredric Remington, and others of the period. I hope you like what I used in the articles as much as I do. Thank you, Charlie and friends, for 'gettin' it all down' for us! All art work used in this book is public domain unless otherwise noted. Some images have been cropped and sized for the page

Thank You . . . !

To my wife, Nancy, who has been my rock for over fifty years – we just celebrated our 50th anniversary. She has put up with me, my wild ideas, my travels and my collecting old junk, and she is also my test reader and edits all my columns before they go out, (and edits and edits and edits and edits.) I could do nothing without you at my side.

To our sons Jeremy and Chad for all you do for us.

To family and friends who read and reread my work and support me no matter how good or bad it might be. My old classmate and friend Phil Singleton went over this manuscript for me, and it was greatly appreciated.

To my friend, editor and mentor, Liane Larocque, for all her help in making it happen.

I can't go without mentioning Working Ranch Magazine and Tim O'Byrne once more. Tim allowed me to take the *Looking Back* column and make it my own. I have loved writing every word of it.

And many thanks to all my wonderful friends from the world of journalism, photography and rodeo, and to the incredible ranch families that took me into their homes, fed me, put up with me and allowed me into their world.

* Referring only to the original Aztec Land and Cattle, no others that came after.

Made in the USA
Columbia, SC
14 August 2019